Studies in Computational Intelligence

Volume 1146

Series Editor

Janusz Kacprzyk, Polish Academy of Sciences, Warsaw, Poland

The series "Studies in Computational Intelligence" (SCI) publishes new developments and advances in the various areas of computational intelligence—quickly and with a high quality. The intent is to cover the theory, applications, and design methods of computational intelligence, as embedded in the fields of engineering, computer science, physics and life sciences, as well as the methodologies behind them. The series contains monographs, lecture notes and edited volumes in computational intelligence spanning the areas of neural networks, connectionist systems, genetic algorithms, evolutionary computation, artificial intelligence, cellular automata, self-organizing systems, soft computing, fuzzy systems, and hybrid intelligent systems. Of particular value to both the contributors and the readership are the short publication timeframe and the world-wide distribution, which enable both wide and rapid dissemination of research output.

Indexed by SCOPUS, DBLP, WTI Frankfurt eG, zbMATH, SCImago.

All books published in the series are submitted for consideration in Web of Science.

Patricia Melin · Oscar Castillo
Editors

New Directions on Hybrid Intelligent Systems Based on Neural Networks, Fuzzy Logic, and Optimization Algorithms

Springer

Editors
Patricia Melin
Division of Graduate Studies and Research
Tijuana Institute of Technology
Tijuana, Mexico

Oscar Castillo
Division of Graduate Studies and Research
Tijuana Institute of Technology
Tijuana, Mexico

ISSN 1860-949X ISSN 1860-9503 (electronic)
Studies in Computational Intelligence
ISBN 978-3-031-53712-7 ISBN 978-3-031-53713-4 (eBook)
https://doi.org/10.1007/978-3-031-53713-4

This Springer imprint is published by the registered company Springer Nature Switzerland AG
The registered company address is: Gewerbestrasse 11, 6330 Cham, Switzerland

Paper in this product is recyclable.

Preface

We describe in this book new directions on the theoretical developments of fuzzy logic, neural networks, and meta-heuristic algorithms, as well as their hybrid combinations, and their application in areas such as intelligent control and robotics, pattern recognition, medical diagnosis, time series prediction, and optimization of complex problems. There are papers with the main topics from type-1 to type-3 fuzzy logic, which basically consists of a group of papers that propose new concepts and algorithms based on type-1, type-2, and type-3 fuzzy logic and their applications. There are also papers that present the theory and practice of meta-heuristics in diverse application areas. There are interesting papers on different applications of fuzzy logic, neural networks, and hybrid intelligent systems in medical problems. In addition, we can find papers describing applications of fuzzy logic, neural networks, and meta-heuristics in robotics problems. Another set of papers presents the theory and practice of neural networks in diverse application areas, including convolutional and deep learning neural networks. There are also a group of papers that present the theory and practice of optimization and evolutionary algorithms in different areas of application. Finally, we can find a group of papers describing the applications of fuzzy logic, neural networks, and meta-heuristics in pattern recognition problems.

In conclusion, the edited book comprises contributions on diverse aspects of fuzzy logic, neural networks, and optimization of meta-heuristic algorithms for designing optimal hybrid intelligent systems and their application in areas such as intelligent control and robotics, pattern recognition, decision-making, time series prediction, and optimization of complex problems. We are hopeful that the book will serve as a guide for researchers and graduate students working in the computational intelligence area.

Tijuana, Mexico
November 2023

Patricia Melin
Oscar Castillo

About This Book

In this book, new directions on the theoretical developments of fuzzy logic, neural networks, and optimization algorithms, as well as their hybrid combinations, are offered. In addition, the above-mentioned methods are discussed in application areas such as control and robotics, pattern recognition, medical diagnosis, decision-making, time series prediction, and optimization of complex problems. Nowadays, the main topics of this book are highly relevant, as most of the intelligent systems and hardware implementations in use manifest some form of intelligent feature to enhance their performance. In addition, theoretically speaking, new and advanced models and algorithms of type-2 and type-3 fuzzy logic are presented, which will be of great interest to researchers in these areas. Also, novel nature-inspired optimization algorithms and innovative neural models are put forward in the manuscript that are very popular subjects at this moment. There are contributions on theoretical aspects as well as applications, which make the book very appealing to a wide audience, ranging from researchers to professors and graduate students.

Contents

Fuzzy Logic

Towards Interval Type-3 Fuzzy Parameter Adaptation in Metaheuristics

Oscar Castillo and Patricia Melin

1 Introduction

It is common knowledge that fuzzy systems can achieve relevant applications in a diverse set of areas. Initially, fuzzy sets were put forward by Lotfi Zadeh in 1965 [1, 2]. Afterwards, fuzzy systems were proposed, and a lot of applications follow, in particular in the control area [3–7].

Parameter adaptation in metaheuristics can be achieved by different methods, including mathematical and probabilistic ones. However, fuzzy logic (type-1) was proposed as another way to achieve parameter adaptation in metaheuristics, such as in [8–10]. Later, type-2 fuzzy logic was also considered for achieving the same task, as a way to enable handling uncertainty due to noise in decision-making, and several algorithms were enhanced in this way [11–14]. However, recently type-3 fuzzy logic has been successfully applied in control [15–19] and other applications [20–23] and we think that the utilizing type-3 fuzzy in parameter adaptation could be even more successful. This is why we put forward the idea of type-3 parameter adaptation and this can be regarded as the contribution of this work.

The article is structured as: Sect. 2 reviews type-3 fuzzy theory, in Sect. 3 we are presenting the method, in Sect. 4 the proposal is highlighted with examples, and in Sect. 6 we are offering conclusions.

O. Castillo (✉) · P. Melin
Tijuana Institute of Technology, TecNM, Tijuana, BC, Mexico
e-mail: ocastillo@tectijuana.mx

P. Melin and O. Castillo (eds.), *New Directions on Hybrid Intelligent Systems Based on Neural Networks, Fuzzy Logic, and Optimization Algorithms*, Studies in Computational Intelligence 1146, https://doi.org/10.1007/978-3-031-53713-4_1

2 Type-3 Fuzzy Theory

Here we are outlining some concepts of type-3 fuzzy theory. Basically, what we propose is changing the fuzzification, inference and defuzzification so that now they are of type-3 form. This, of course, will allow having more ability to handle uncertainty. The fuzzy rules should be the same, as the knowledge for control is the same, the changes occur in the geometry of the membership functions (MFs) and the inference process. For more details regarding type-3 please check the references [20–22]. Just to illustrate the difference in the form of type-3 MF, please check in Fig. 1 a sample function of this type. In this figure we illustrate the footprint of uncertainty (FOU), which is limited the lower and upper MFs indicated with "f" in the image.

Definition 1 A type-3 fuzzy set (T3 FS) [20], denoted by $A^{(3)}$, is represented by a function, MF of $A^{(3)}$, in the Cartesian product $X \times [0, 1] \times [0, 1]$ in [0, 1], where X is the primary universe of $A^{(3)}$, x. The MF of $\mu_{A^{(3)}}$ is formulated by $\mu_{A^{(3)}}(x, u, v)$ (or $\mu_{A^{(3)}}$ for short) and it is labeled a type-3 MF (T3 MF). In other words,

$$\mu_{A^{(3)}} : X \times [0, 1] \times [0, 1] \rightarrow [0, 1]$$

$$A^{(3)} = \{(x, u(x), v(x, u), \mu_{A^{(3)}}(x, u, v))|x \in X, u \in U \subseteq [0, 1], v \in V \subseteq [0, 1]\} \quad (1)$$

where U is the universe for the secondary variable u and V is the universe for tertiary variable v.

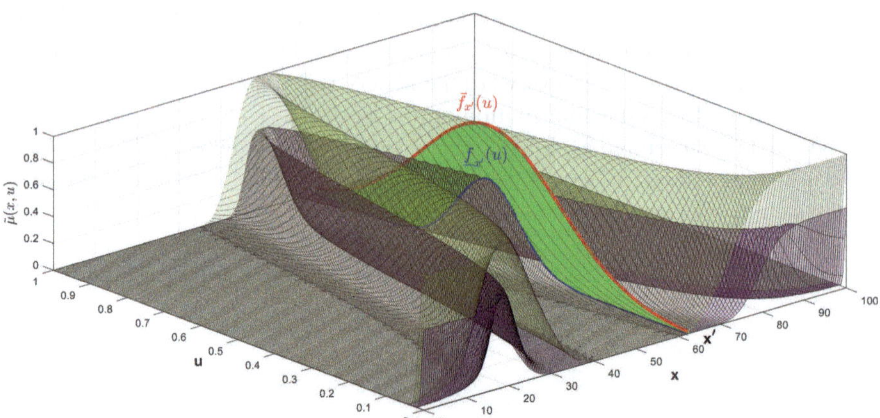

Fig. 1 Interval type-3 membership function

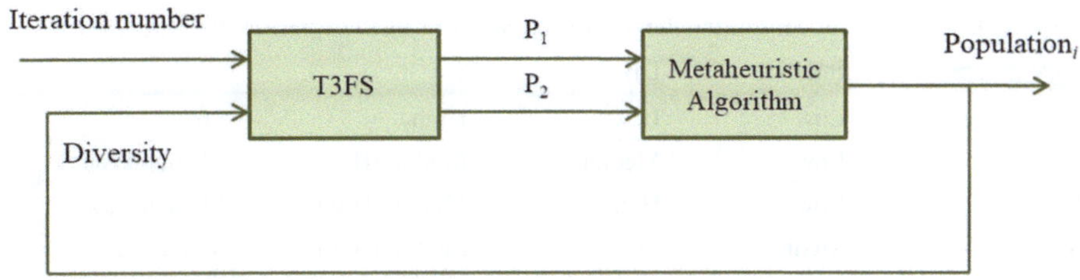

Fig. 2 Type-3 fuzzy parameter adaptation approach

3 Type-3 Fuzzy Adaptation

Here the proposed type-3 fuzzy parameter adaptation in metaheuristics is presented. Figure 2 illustrates with a block diagram the idea of the approach. A type-3 fuzzy system (T3FS) is employed to estimate the parameter values (P_1 and P_2) of the metaheuristic a particular iteration number (i) and the result is a new population of individuals. This population at iteration i is fed back to the T3FS and the diversity metric can be calculated to estimate the state of the convergence (in the metaheuristic). This is repeated for a number of iterations or when the desired accuracy is achieved. Of course, the approach shown in Fig. 2 could be generalized to consider more parameter (not just two) and more metrics for the metaheuristic could be utilized (not just the diversity of the individuals).

For example, for a genetic algorithm, P_1 could be the crossover rate and P_2 could be mutation rate.

4 Illustrative Examples

In general, all metaheuristic algorithms have several parameters that have an effect on the accuracy and rate of convergence. If we assume that only two parameters are utilized in the algorithm then the situation of Fig. 2 represents the parameterization problem. For example, in particle swarm optimization (PSO), the parameters could be as follows: P_1 = social coefficient, and P_2 = cognitive coefficient. Another example could be the bee colony optimization (BCO), where we can define the parameters as $P_1 = \alpha$ and $P_2 = \beta$. We can also mention the differential evolution (DE) algorithm, where we can use: P_1 = crossover rate, and P_2 = mutation rate (F). Of course, more examples could be mentioned, but the idea is similar.

For BCO the fuzzy system for controlling the parameter values is shown in Table 1. The fuzzy system inputs are the "iteration" and "diversity". The outputs are the values of the alpha and beta parameters of BCO, please check more details in, where this was done utilizing type-2 [11]. The rules are the same when we extend to type-3, just the MFs will now be type-3, as in Fig. 1.

Table 1 Knowledge for controlling the parameter values in BCO expressed with fuzzy rules

Rule number	Iteration	Diversity	β	α
1	Low	Low	High	Low
2	Low	Medium	MediumHigh	MediumLow
3	Low	High	MediumHigh	MediumLow
4	Medium	Low	MediumHigh	MediumLow
5	Medium	Medium	Medium	Medium
6	Medium	High	MediumLow	MediumHigh
7	High	Low	Medium	High
8	High	Medium	MediumLow	MediumHigh
9	High	High	Low	High

The linguistic values in Table 1 are represented by type-3 fuzzy sets [20], which need an appropriate parameterization to achieve the best possible results.

5 Conclusions

An approach for fuzzy parameter adaptation in metaheuristics using type-3 fuzzy logic has been presented. Previous works in this area have shown the success of utilizing type-2 fuzzy logic for adaptation in several metaheuristics, such as harmony search, cuckoo search, differential evolution and others. However, an extension for using type-3 is proposed with the idea of handling in a better way the intrinsic uncertainty in solving search and optimization problems with metaheuristics. The idea resides in the utilization of a type-3 system for obtaining the main parameter values of a particular metaheuristic, during execution, with the goal of enhancing convergence and accuracy of the algorithm. The proposed approach is illustrated with several algorithms, but could be extended to other ones in the future, like for the algorithms mentioned in [10–14] or in [24, 25]. Finally, the optimization approach could be tested with problems in diverse areas, like [26–28].

References

1. L.A. Zadeh, Fuzzy sets. Inf. Control. **8**, 338–353 (1965)
2. L.A. Zadeh, Knowledge representation in fuzzy logic. IEEE Trans. Knowl. Data Eng. **1**, 89 (1989)
3. O. Montiel, R. Sepulveda, P. Melin, O. Castillo, M.A. Porta García, I.A., Meza-Sánchez, Performance of a simple tuned fuzzy controller and a PID controller on a DC motor. FOCI 531–537 (2007)
4. P. Melin, O. Castillo, A new method for adaptive control of non-linear plants using type-2 fuzzy logic and neural networks. Int. J. Gen. Syst. **33**(2–3), 289–304 (2004)

5. T. Kumbasar, H. Hagras, in *Interval Type-2 Fuzzy PID Controllers*, eds. by J. Kacprzyk, W. Pedrycz, Springer Handbook of Computational Intelligence (Springer Handbooks, Springer, Berlin, Heidelberg, 2015)
6. O. Castillo, P. Melin, A new fuzzy-fractal-genetic method for automated mathematical modelling and simulation of robotic dynamic systems, 1998 IEEE International Conference on Fuzzy Systems (FUZZ-IEEE 1998) Proceedings. Vol. 2, 1182–1187
7. O. Castillo, P. Melin, Intelligent adaptive model-based control of robotic dynamic systems with a hybrid fuzzy-neural approach. Appl. Soft Comput. **3**(4), 363–378 (2003)
8. F. Valdez, P. Melin, O. Castillo, Evolutionary method combining particle swarm optimization and genetic algorithms using fuzzy logic for decision making, in: IEEE International Conference on Fuzzy Systems, 2009, pp. 2114–2119
9. F. Valdez, J.C. Vazquez, P. Melin, O. Castillo, Comparative study of the use of fuzzy logic in improving particle swarm optimization variants for mathematical functions using co-evolution. Appl. Soft Comput. Comput. **52**, 1070–1083 (2017)
10. D. Sanchez, P. Melin, O. Castillo, A grey wolf optimizer for modular granular neural networks for human recognition. Comput. Intell. Neurosci. (2017).https://doi.org/10.1155/2017/4180510
11. O. Castillo, L. Amador-Angulo, A generalized type-2 fuzzy logic approach for dynamic parameter adaptation in bee colony optimization applied to fuzzy controller design. Inf. Sci. **460**, 476–496 (2018)
12. J. Perez, F. Valdez, O. Castillo, O. Roeva, Bat algorithm with parameter adaptation using interval type-2 fuzzy logic for benchmark mathematical functions. In 2016 IEEE 8th international Conference on Intelligent Systems (IS) (IEEE, 2016), pp. 120–127
13. F. Cuevas, O. Castillo, P. Cortés-Antonio, Generalized type-2 fuzzy parameter adaptation in the marine predator algorithm for fuzzy controller parameterization in mobile robots. Symmetry **14**, 859 (2022). https://doi.org/10.3390/SYM14050859
14. M. Guerrero, F. Valdez, O. Castillo, Comparative study between type-1 and interval type-2 fuzzy systems in parameter adaptation for the cuckoo search algorithm. Symmetry **14**, 2289 (2022). https://doi.org/10.3390/SYM14112289.
15. S.N. Qasem, A. Ahmadian, A. Mohammadzadeh, S. Rathinasamy, B. Pahlevanzadeh, A type-3 logic fuzzy system: optimized by a correntropy based Kalman filter with adaptive fuzzy kernel size. Inform. Sci. **572**, 424–443 (2021)
16. A. Mohammadzadeh, M.H. Sabzalian, W. Zhang, An interval type-3 fuzzy system and a new online fractional-order learning algorithm: theory and practice. IEEE Trans. Fuzzy Syst. **28**(9), 1940–1950 (2020)
17. Z. Liu, A. Mohammadzadeh, H. Turabieh, M. Mafarja, S.S. Band, A. Mosavi, A new online learned interval type-3 fuzzy control system for solar energy management systems. IEEE Access **9**, 10498–10508 (2021)
18. J.H. Wang, J. Tavoosi, A. Mohammadzadeh, S. Mobayen, J.H. Asad, W. Assawinchaichote, P. Skruch, Non-singleton type-3 fuzzy approach for flowmeter fault detection: experimental study in a gas industry. Sensors **21**(21), 7419 (2021)
19. K.A. Alattas, A. Mohammadzadeh, S. Mobayen, A.A. Aly, B.F. Felemban, A new data-driven control system for memss gyroscopes: dynamics estimation by type-3 fuzzy systems. Micromachines **12**(11), 1390 (2021)
20. O. Castillo, J.R. Castro, P. Melin, *Interval Type-3 Fuzzy Systems: Theory and Design* (Springer, Cham, Switzerland, 2022)
21. J.T. Rickard, J. Aisbett, G. Gibbon, Fuzzy subsethood for fuzzy sets of type-2 and generalized type-n. IEEE Trans. Fuzzy Syst. **17**(1), 50–60 (2009)
22. O. Castillo, Towards finding the optimal *n* in designing type-*n* fuzzy systems for particular classes of problems: a review. Appl. Comput. Math. **17**(1), 3–9 (2018)
23. O. Castillo, J.R. Castro, P. Melin, Interval type-3 fuzzy control for automated tuning of image quality in televisions. Axioms **11**, 276 (2022)
24. O. Castillo, E. Lizzarraga, J. Soria, P. Melin, F. Valdez, New approach using [24] ant colony optimization with ant set partition for fuzzy control design applied to the ball and beam system. Inf. Sci. **294**, 203–215 (2015)

25. L. Amador-Angulo, O. Mendoza, J.R. Castro, A. Rodriguez-Diaz, P. Melin, O. Castillo, Fuzzy sets in dynamic adaptation of parameters of a bee colony optimization for controlling the trajectory of an autonomous mobile robot. Sensors **16**(9), 1458 (2016)
26. E. Ontiveros, P. Melin, O. Castillo, Comparative study of interval type-2 and general type-2 fuzzy systems in medical diagnosis. Inf. Sci. **525**, 37–53 (2020)
27. J.E. Moreno, M.A. Sanchez, O. Mendoza, A. Rodriguez-Diaz, O. Castillo, P. Melin, J.R. Castro, Design of an interval type-2 fuzzy model with justifiable uncertainty. Inf. Sci. **513**, 206–221 (2020)
28. B. González, F. Valdez, P. Melin, G. Prado-Arechiga, Fuzzy logic in the gravitational search algorithm for the optimization of modular neural networks in pattern recognition. Expert Syst. Appl. **42**(14), 5839–5847 (2015)

An Application of Controlled Sets in Medical Diagnosis

Sinem Tarsuslu (Yılmaz) and Gökhan Çuvalcıoğlu

1 Introduction

Fuzzy set theory, firstly introduced by Zadeh [13], served as an expansion of crisp sets, aiming to address the limitations of binary logic and facilitate the representation of uncertainty and vagueness in decision-making processes. This innovative concept allowed for the inclusion of membership degree, enabling sets to contain elements with degrees of membership rather than strict inclusion or exclusion.

Building upon Zadeh's fuzzy set theory, the intuitionistic fuzzy set concept was introduced by Atanassov [1]. Atanassov extended the fuzzy sets by broadening the truth value set into a lattice $L^* = \{(x_1, x_2) \in [0, 1]^2 : x_1+x_2 \leq 1\} \subset [0, 1]\times[0, 1]$. This expansion enabled the consideration of not only membership degrees but also non-membership degrees and a separate parameter called hesitation, enhancing the ability to model uncertainty in a more nuanced manner. The inclusion of these additional parameters provided a more comprehensive representation of imprecision and uncertainty in decision-making and problem-solving scenarios. Intuitionistic fuzzy sets are employed in logical programming, machine learning, and decision-making problem-solving arenas. In modeling decision-making problems, the hesitation degree of intuitionistic fuzzy sets have an important role. High success rates have been obtained in its application to decision-making contexts. One of the most important of these is the medical diagnosis application.

Various approaches were implemented for medical diagnosis modelling. Sanchez defined the concept of medical knowledge with fuzzy relations [8]. Then, De and

S. Tarsuslu (Yılmaz) (✉)
Department of Natural and Mathematical Sciences, Faculty of Engineering, Tarsus University, 33400 Tarsus, Turkey
e-mail: sinemtarsuslu@tarsus.edu.tr

G. Çuvalcıoğlu
Department of Mathematics, Faculty of Sciences, Mersin University, Mersin, Turkey

P. Melin and O. Castillo (eds.), *New Directions on Hybrid Intelligent Systems Based on Neural Networks, Fuzzy Logic, and Optimization Algorithms*, Studies in Computational Intelligence 1146, https://doi.org/10.1007/978-3-031-53713-4_2

his friends extended this approach to intuitionistic fuzzy sets. Szmidt and Kacprzyk introduced a new intuitionistic fuzzy method using the smallest distance between symptoms. Also, many authors have studied applications of medical diagnosis in intuitionistic fuzzy sets with different approaches, as composition of relations, distance measure, similarity measure and several different operators [7, 9–11]. In order to obtain the most accurate diagnosis, researchers continue to study widely.

In applications of intuitionistic fuzzy sets, the degree of the non-membership degree of the criteria is defined independently of the degree of membership. There is generally no method for determining the degree of non-membership. Since this situation directly affects the hesitation degree, it negatively affects to get decision. The theory of controlled sets was defined by Çuvalcoğlu [3]. Çuvalcoğlu's research demonstrated that the controlled set can be regarded as a specialized instance of the intuitionistic fuzzy set. However, despite this relationship, the controlled set differs significantly from the intuitionistic fuzzy set, primarily in the determination of the non-membership degree. In this theory, the non-membership degree of an element is closely intertwined to the membership degrees of all elements within the set.

Çuvalcoğlu extensively investigated the operations, fundamental algebraic properties, and topological aspects specific to controlled sets. This comprehensive study contributed to a deeper understanding of the characteristics and behaviors of controlled sets within mathematical frameworks. Moreover, beyond the theoretical groundwork, applications of controlled sets were explored across diverse domains [5, 12]. These applications span various fields, showcasing the practical relevance and utility of the controlled set theory in real-world scenarios. Researchers have investigated and implemented controlled sets in different areas, leveraging their unique properties to address complex problems and enhance decision-making processes across multiple disciplines.

In the present paper, we studied the approach introduced by Szmidt and Kacprzyk using the Controlled Set theory. We reconstructed the tables giving the patient-symptom and disease-symptom relations using the controlled set definition. And in the final we got the diagnosis using the normalized Hamming distance and normalized Euclidean distance.

Firstly, we will introduce essential definitions pivotal to this paper.

Definition 1 [1] An intuitionistic fuzzy set (IFS) on a set X is an object.

$$A = \{\langle x,\ \mu_A(x),\ \nu_A(x)\rangle \ :\ x \in X\}$$

where $\mu_A(x)$, ($\mu_A : X \to [0, 1]$) is the degree of membership of x in A, $\nu_A(x)$, ($\nu_A : X \to [0, 1]$) is the degree of non- membership of x in A and μ_A and ν_A satisfy the following condition:

$$\mu_A(x) + \nu_A(x) \leq 1, \text{ for all } x \in X.$$

The class of intuitionistic fuzzy sets on X is denoted by $IFS(X)$.

The hesitation degree of x is $\pi_A(x) = 1 - \mu_A(x) - \nu_A(x)$.

Basic operations on intuitionistic fuzzy sets are given as follow;

Definition 2 [1] An IFS A is said to be contained in an IFS B ($A \sqsubseteq B$) if and only if, for all $x \in X \; : \; \mu_A(x) \leq \mu_B(x)$ and $\nu_A(x) \geq \nu_B(x)$.

$A = B \Leftrightarrow A \sqsubseteq B$ and $B \sqsubseteq A$.

Definition 3 [1] Let $A \in IFS$ and $A = \{\langle x, \mu_A(x), \nu_A(x)\rangle \; : \; x \in X\}$ then the complement of A is.

$$A^c = \{\langle x, \nu_A(x), \mu_A(x)\rangle \; : \; x \in X\}.$$

The intersection and the union of two IFSs A and B on X is defined by

$$A \sqcap B = \{\langle x, \min\{\mu_A(x), \mu_B(x)\}, \max\{\nu_A(x), \nu_B(x)\}\rangle : \; x \in X\}$$

$$A \sqcup B = \{\langle x, \max\{\mu_A(x), \mu_B(x)\}, \min\{\nu_A(x), \nu_B(x)\}\rangle \; : \; x \in X\}$$

Some special Intuitionistic Fuzzy Sets on X are defined as following;

$$O^* = \{\langle x, 0, 1\rangle \; : \; x \in X\}$$

$$X^* = \{< x, 1, 0 >: \; x \in X\}$$

Definition 4 [8] Intuitionistic fuzzy relation (IFR) is an intuitionistic fuzzy subset of $X \times Y$ that is, is an expression R given by.

$$R = \{\langle (x, y), \mu_R(x, y), \nu_R(x, y)\rangle \; : \; x \in X, \; y \in Y\}$$

with $0 \leq \mu_R(x, y) + \nu_R(x, y) \leq 1$ for any $(x, y) \in X \times Y$. It will be denoted by $R(X \to Y)$.

Definition 5 [8] Let $R_1 \in IFR(X \times Y)$ and $R_2 \in IFR(Y \times Z)$ are two IFRs. The composition $R_1 \circ R_2$ is the intuitionistic fuzzy relation from X to Z, defined by the membership function.

$$\mu_{R_1 \circ R_2}(x, z) = \bigvee_y \left[\mu_{R_1}(x, y) \wedge \mu_{R_2}(y, z)\right]$$

and the non-membership function

$$\nu_{R_1 \circ R_2}(x, z) = \bigwedge_y \left[\nu_{R_1}(x, y) \wedge \nu_{R_2}(y, z)\right]$$

$\forall (x, z) \in X \times Z$ and $\forall y \in Y$.

Now, the definition and basic algebraic properties of the controlled set will be given.

Definition 6 [3] Let U be a universal, $\alpha : U \to I$ is called α-set.

From the definition, if X is a universal and μ is a fuzzy set on X then X is a $\mu-$ set.

Definition 7 [3] Let U be an α-set. The set U is called α-controlled set if.

$\forall x \in U, \exists y \in U \ni 1 - \alpha(x) = \alpha(y)$.

If U is α-controlled set then it is shown by $U \in CS(\alpha)$.

Definition 8 [4] Let $U \in CS(\alpha)$ and $a \in U$. The control set of a is.

$\overline{a} = \{b \in E \ : \ 1 - \mu(a) = \mu(b)\}$.

Theorem 1 [4] Let $U \in CS(\alpha)$ and $a,\ b \in U$. Then;

(1) $a \in \overline{a} \Leftrightarrow \mu(a) = 1/2$
(2) $a \in \overline{b} \Leftrightarrow b \in \overline{a}$
(3) If $\overline{a} \neq \overline{b}$ then $x \in \overline{a} \Leftrightarrow x \notin \overline{b}$
(4) If $\overline{a} \cap \overline{b} \neq \emptyset$ then $\overline{a} = \overline{b}$

This theorem implies a straightforward conclusion: it's impossible to establish an equivalence relation on the universal set U in such a way that the control sets of the elements form equivalence classes. This impossibility stems from the fact that specific elements might not inherently belong to their respective classes within this context. However, in the context of control sets, the determination of an element's membership within a class is influenced by a broader range of factors, including degrees of non-membership and the interrelatedness of elements within the set.

Proposition 1 [3] Let U be an α-set and $C = \{X \subset U \ : \ X \in CS(\alpha)\}$. Then,

i. $\bigcup\limits_{X \varepsilon C' \subset C} X \in CS(\alpha)$
ii. The chain of α-controlled sets $X_1 \subset X_2 \subset X_3 \subset \cdots$ is finite.
iii. The class C have maximal element property.

The significance of the membership degree within a set for an element is undeniable. However, the non-membership degree holds equal importance. Not all sets can be categorized as controlled sets. Nonetheless, by utilizing the membership degrees of elements, we can introduce the concept of controlled set. Hence, we can give the following definition:

Definition 9 [4] Let U be α-set. We define the following mapping on U such that.

$$\alpha^*(x) = \begin{cases} 1 - \alpha(x), & x \in U_\alpha \\ \sup\limits_y \alpha(y), & y \in U \ni \alpha(x) < 1 - \alpha(y) \\ 0, & \text{Otherwise.} \end{cases}$$

where $U_\alpha = \bigcup\limits_{a \in U} \overline{a}$.

It is evident that α^* is a mapping from U to I. Moreover, it is readily observable that the sum of α^* and α is less than or equal to 1. From this properties, we can give the following definition;

Definition 10 [4] Let U be α-set. Then the set $A = \{< x,\ \alpha(x),\ \alpha^*(x) >:\ \ x \in U\}$ is called $(\alpha,\ \alpha^*)-$ controlled set.

From the definition, it can be easily seen that every $(\alpha,\ \alpha^*)-$ controlled set is intuitionistic fuzzy set. But the converse of this is not true generaly.

2 Application of Intuitionistic Fuzzy Sets to Medical Diagnosis

The realm of medicine and decision-making stands out as one of the most promising and captivating domains where fuzzy set theory finds its relevance. In practical scenarios, the inherent vagueness within medical records and the uncertain data pivotal for decision-making necessitate the utilization of fuzzy logic. Since the inception of fuzzy set theory, numerous expansions have emerged and found effective application in the field of medical diagnosis, demonstrating its efficacy in handling imprecise information and improving decision-making processes.

In 2001 [6], De and colleagues introduced an application of intuitionistic fuzzy sets for medical diagnosis, expanding on Sanchez's approach [8]. They outlined the methodology through three primary steps:

(1) Identifying and determining symptoms
(2) Formulating a medical knowledge base utilizing intuitionistic fuzzy relations
(3) Determining diagnoses based on the composition of intuitionistic fuzzy relations

In the context of medical analysis, a group consisting of patients $p_{i,}i = 1, ..., n$, denoted as P, was studied. Every patient was linked to a distinct collection of symptoms denoted as S. An intuitionistic fuzzy relation, Q, was established to correlate the set of patients (P) with their respective symptoms (S). Following this, a second intuitionistic fuzzy relation, R, was devised to map the set of symptoms (S) to a range of possible diagnoses (D). The combined composition, T, of these intuitionistic fuzzy relations R and Q aimed to define and describe the condition of patients in relation to their diagnosis. This process served to connect patient symptoms with potential diagnoses, thereby establishing a bridge that aids in understanding and characterizing the patients' conditions.

For the outcome analysis, a calculation was performed on T, $S_R = \mu_R - \nu_R \pi_R$. This computation generated a probable diagnosis table, enabling the determination of the diagnosis by identifying the maximum value among the calculated results.

Szmidt and Kacprzyk introduced a new methodology for addressing the same medical diagnostic problem, criticizing the limitations of the previously employed method. Their proposed approach involved seeking the minimal distance between symptoms as a means of diagnosing patients [10]. In their study, two distinct distance functions were utilized for computation: the normalized Hamming distance and the normalized Euclidean distance. These distance metrics served as crucial factors in determining the relationships and similarities between symptoms, contributing to a different perspective on the diagnosis process compared to the previous method.

Example 1 [6, 10] Let there be four patients: Al, Bob, Joe and Ted, i.e. $P = \{Al, Bob, Joe, Ted\}$. The set of symptoms under consideration is denoted by the $S = \{temperature, headache, stomach\ pain, cough, chest - pain\}$. The intuitionistic fuzzy relation $Q(P \to S)$ is in Table 1. Let the set of diagnoses be $D = \{Viral\ fever, Malaria, Typhoid, Stomach, Problem, Chest\ Problem\}$. The intuitionistic fuzzy relation $R(S \to D)$ is in Table 2.

The normalised Hamming distance for all the symptoms exhibited by the $i - th$ patient and the $k - th$ diagnosis is calculated as

$$l(s(p_i), d_k) = \frac{1}{10} \sum_{j=1}^{5} \begin{pmatrix} |\mu_j(p_i) - \mu_j(d_k)| + |\nu_j(p_i) - \nu_j(d_k)| \\ +|\pi_j(p_i) - \pi_j(d_k)| \end{pmatrix}$$

The distances of each patient from the specified set of potential diagnoses are provided in Tables 3 and 4: Al is afflicted with Malaria, Bob with Stomach Problems, Joe with Typhoid, and Ted with Fever.

They obtained same quality diagnosis for each patient by applying the normalized Euclidean distance:

Table 1 The intuitionistic fuzzy relation $Q(P \to S)$

	Temperature	Headache	Stomach pain	Cough	Chest pain
Al	(0.8,0.1,0.1)	(0.6,0.1,0.3)	(0.2,0.8,0.0)	(0.6,0.1,0.3)	(0.1,0.6,0.3)
Bob	(0.0,0.8,0.2)	(0.4,0.4,0.2)	(0.6,0.1,0.3)	(0.1,0.7,0.2)	(0.1,0.8,0.1)
Joe	(0.8,0.1,0.1)	(0.8,0.1,0.1)	(0.0,0.6,0.4)	(0.2,0.7,0.1)	(0.0,0.5,0.5)
Ted	(0.6,0.1,0.3)	(0.5,0.4,0.1)	(0.3,0.4,0.3)	(0.7,0.2,0.1)	(0.3,0.4,0.3)

Table 2 The intuitionistic fuzzy relation $R(S \to D)$

	Viral fever	Malaria	Typhoid	Stomach problem	Chest problem
Temperature	(0.4,0.0,0.6)	(0.7,0.0,0.3)	(0.3,0.3,0.4)	(0.1,0.7,0.2)	(0.1,0.8,0.1)
Headache	(0.3,0.5,0.2)	(0.2,0.6,0.2)	(0.6,0.1,0.3)	(0.2,0.4,0.4)	(0.0,0.8,0.2)
Stomach pain	(0.1,0.7,0.2)	(0.0,0.9,0.1)	(0.2,0.7,0.1)	(0.8,0.0,0.2)	(0.2,0.8,0.0)
Cough	(0.4,0.3,0.3)	(0.7,0.0,0.3)	(0.2,0.6,0.2)	(0.2,0.7,0.1)	(0.2,0.8,0.0)
Chest pain	(0.1,0.7,0.2)	(0.1,0.8,0.1)	(0.1,0.9,0.0)	(0.2,0.7,0.1)	(0.8,0.1,0.1)

Table 3 The normalized Hamming distances for each patient

	Viral Fever	Malaria	Typhoid	Stomach problem	Chest problem
Al	0.28	0.24	0.28	0.54	0.56
Bob	0.40	0.50	0.31	0.14	0.42
Joe	0.38	0.44	0.32	0.50	0.55
Ted	0.28	0.30	0.38	0.44	0.54

Table 4 The normalized Euclidean distances for each patient

	Viral fever	Malaria	Typhoid	Stomach problem	Chest problem
Al	0.29	0.25	0.32	0.53	0.58
Bob	0.43	0.56	0.33	0.14	0.46
Joe	0.36	0.41	0.32	0.52	0.57
Ted	0.25	0.29	0.35	0.43	0.50

$$q(s(p_i), d_k) = \left(\frac{1}{10} \sum_{j=1}^{5} \left(\mu_j(p_i) - \mu_j(d_k)\right)^2 + \left(\nu_j(p_i) - \nu_j(d_k)\right)^2 + \left(\pi_j(p_i) - \pi_j(d_k)\right)^2 \right)^{\frac{1}{2}}.$$

3 Main Results

In this paper, our purpose is to examine the same problem with controlled set theory and discuss the results. From this point of view, firstly, intuitionistic fuzzy relations Q and R redefined by controlled set theory as in Tables 5 and 6. The new relations obtained are controlled relations so that the value $\nu(x)$ is now defined as $\mu^*(x)$. The values derived from both the normalized Hamming distance and the normalized Euclidean distance demonstrated consistent relationships.

Table 5 The controlled relation $Q(P \to S)$

	Temperature	Headache	Stomach pain	Cough	Chest pain
Al	(0.8,0.2,0.0)	(0.6,0.4,0.0)	(0.2,0.8,0.0)	(0.6,0.4,0.0)	(0.1,0.8,0.1)
Bob	(0.0,0.8,0.2)	(0.4,0.6,0.0)	(0.6,0.4,0.0)	(0.1,0.8,0.1)	(0.1,0.8,0.1)
Joe	(0.8,0.2,0.0)	(0.8,0.2,0.0)	(0.0,0.8,0.2)	(0.2,0.8,0.0)	(0.0,0.8,0.2)
Ted	(0.6,0.4,0.0)	(0.5,0.5,0.0)	(0.3,0.7,0.0)	(0.7,0.3,0.0)	(0.3,0.7,0.0)

Table 6 The controlled relation $R(S \to D)$

	Viral fever	Malaria	Typhoid	Stomach problem	Chest problem
Temperature	(0.4,0.6,0.0)	(0.7,0.3,0.0)	(0.3,0.7,0.0)	(0.1,0.8,0.1)	(0.1,0.8,0.1)
Headache	(0.3,0.7,0.0)	(0.2,0.8,0.0)	(0.6,0.4,0.0)	(0.2,0.8,0.0)	(0.0,0.8,0.2)
Stomach pain	(0.1,0.8,0.1)	(0.0,0.8,0.2)	(0.2,0.8,0.0)	(0.8,0.2,0.0)	(0.2,0.8,0.0)
Cough	(0.4,0.6,0.0)	(0.7,0.3,0.0)	(0.2,0.8,0.0)	(0.2,0.8,0.0)	(0.2,0.8,0.0)
Chest pain	(0.1,0.8,0.1)	(0.1,0.8,0.1)	(0.1,0.8,0.1)	(0.2,0.8,0.0)	(0.8,0.2,0.0)

Table 7 Applying the normalized Hamming distance

	Viral fever	Malaria	Typhoid	Stomach problem	Chest problem
Al	0.20	0.16	0.18	0.44	0.48
Bob	0.26	0.42	0.20	0.14	0.34
Joe	0.26	0.26	0.20	0.46	0.50
Ted	0.18	0.26	0.16	0.30	0.34

Table 8 Applying the normalized Euclidean distance

	Viral fever	Malaria	Typhoid	Stomach problem	Chest problem
Al	0.24	0.21	0.29	0.47	0.52
Bob	0.29	0.44	0.24	0.15	0.40
Joe	0.31	0.35	0.26	0.52	0.55
Ted	0.17	0.27	0.17	0.34	0.37

Table 7 is obtained by applying the normalized Hamming distance and Table 8 is by applying the normalized Euclidean distance to the $(\mu, \mu^*)-$ controlled set formed with $\mu-$ sets. The appropriate diagnoses derived from Tables 7 and 8 are as follows: Al is diagnosed with Malaria, Bob with Stomach Problems, while both Joe and Ted are diagnosed with Typhoid.

4 Conclusion

In the intuitionistic fuzzy medical diagnosis, several methodologies have been devised to achieve optimal diagnoses, employing diverse approaches such as distance functions, similarity functions, aggregation operators, modal operators, and more. The focus of our study revolves around the smallest distance approach, as introduced by Szmidt and Kacprzyk. Consider the case of Ted, where the proper diagnosis, based on conventional methods, suggests Fever. However, an intriguing discrepancy arises: while Ted's reported membership degree for a headache concerning Fever remains below 0.5, Ted's complaint of a headache stands at 0.5. Additionally, although stomach pain isn't a defining symptom for Fever, Ted's reported stomach complaint holds a substantial degree of membership. These inconsistencies hint at the possibility of a different diagnosis for Ted. In contrast, when employing controlled sets and employing normalized Hamming and normalized Euclidean distances in our analysis, Ted's diagnosis shifts to Typhoid. Notably, in Typhoid, a headache emerges as a distinctive symptom, aligning with Ted's reported complaint. Furthermore, fever stands as the second most influential symptom. This revised evaluation using controlled sets enables a more nuanced and sensitive assessment, facilitating the differentiation between diseases that exhibit similar symptoms. This approach illustrates the significance of controlled sets in refining medical diagnoses, allowing for a more detailed and accurate distinction among diseases with overlapping symptomatology. By integrating these methodologies, we can enhance diagnostic precision and better address cases where symptoms overlap or diverge from conventional patterns.

References

1. K.T. Atanassov, *Intuitionistic Fuzzy Sets*, VII ITKR.s Session, Sofia, (1983)
2. K.T. Atanassov, *Studies in Fuzziness and Soft Computing-On Intuitionistic Fuzzy Sets Theory, ISBN 978-3-642-29126-5* (Springer, Heidelberg, New York, 2012)
3. G. Çuvalcoğlu, *Controlled Set Theory, Bogolyubov Readings DIF-2013.* (Ukraine, 2013), p. 342
4. G. Çuvalcoğlu, Some properties of controlled set theory. Notes IFSs **20**(2), 37–42 (2014)
5. G. Çuvalcoğlu, S.U. Varol, Decision making process for serving restaurants using intuitionistic fuzzy set theory via controlled sets. J. Univ. Math. **4**(2), 296–325 (2021)
6. S.K. De, R. Biswas, A.R. Roy, An application of intuitionistic fuzzy sets in medical diagnosis. Fuzzy Sets Syst. **117**, 209–213 (2001)
7. M. Luo, R. Zhao, A distance measure between intuitionistic fuzzy sets and its application in medical diagnosis. Artif. Intell. Med. **89**, 34–39 (2018)
8. E. Sanchez, in *Solutions in Composite Fuzzy Relation Equation. Application to Medical Diagnosis in Brouwerian Logic*, ed. by M.M. Gupta, G.N. Saridis, B.R. Gaines Fuzzy Automata and Decision Process (Elsevier, North-Holland, 1977)
9. Y. Song, X. Wang, W. Quan, W. Huang, A new approach to construct similarity measure for intuitionistic fuzzy sets. Soft. Comput. **23**, 1985–1998 (2019)
10. E. Szmidt, J. Kacprzyk, Distances between intuitionistic fuzzy sets. Fuzzy Sets Syst. **114**(3), 505–518 (2000)
11. E. Szmidt, J. Kacprzyk, Intuitionistic fuzzy sets in some medical applications, Fifth Int. Conf. on IFSs, Sofia, 22–23 Sept., NIFS **7**(4), 58–64 (2001).
12. F. Tuğrul, Application of intuitionistic fuzzy logic with a new method in multi criteria decision making process. KSU, Department of Mathematics, PhD. Thesis, pp. 199 (2021)
13. L.A. Zadeh, Fuzzy sets. Inf. Control **8**, 338–353 (1965)

Dynamic Adjustment of an Artificial Fish Swarm Algorithm Using Shadowed Type-2 Fuzzy Logic for Enhancing Benchmark Function Optimization

Patricia Ochoa, Cinthia Peraza, and Oscar Castillo

1 Introduction

Fuzzy logic, a revolutionary concept in mathematics and computer science, has significantly transformed various fields by introducing a nuanced approach to dealing with uncertainty and imprecision. This concept, proposed by Lotfi Zadeh in the 1960s, departs from classical binary logic and acknowledges the existence of partial truths, allowing for a more realistic representation of complex and uncertain information [1].

Fuzzy logic operates on the principle of "degrees of truth," assigning values between 0 and 1 to statements, accommodating situations where an assertion can be partially true and partially false simultaneously. This flexibility is particularly valuable when addressing real-world problems that involve ambiguity or vagueness. Applications span across diverse domains including control systems [2–4], decision-making [5], pattern recognition [6, 7], and more.

In recent years, advances in fuzzy logic have boosted its usefulness and sophistication. The integration of fuzzy logic with other technologies, such as machine learning and data analytics, has improved problem-solving capabilities. Hybrid models that combine fuzzy logic with neural networks or genetic algorithms have shown exceptional results in complex tasks such as medical diagnosis, image processing, and financial forecasting.

P. Ochoa · C. Peraza · O. Castillo (✉)
Division of Graduate Studies, Tijuana Institute of Technology, TecNM, Tijuana, México
e-mail: ocastillo@tectijuana.mx

P. Ochoa
e-mail: martha.ochoa18@tectijuana.edu.mx

C. Peraza
e-mail: cinthia.peraza@tectijuana.edu.mx

P. Melin and O. Castillo (eds.), *New Directions on Hybrid Intelligent Systems Based on Neural Networks, Fuzzy Logic, and Optimization Algorithms*, Studies in Computational Intelligence 1146, https://doi.org/10.1007/978-3-031-53713-4_3

In the realm of decision-making, fuzzy logic has enabled the development of expert systems that can emulate human decision processes. These systems excel in scenarios where explicit rule-based reasoning is challenging due to the complexity and imprecision of the data involved.

Furthermore, type-2 fuzzy logic has emerged as a more potent extension of the conventional type-1 fuzzy logic. This extension deals with uncertainty in a more nuanced manner by incorporating a "fuzzy set of fuzzy sets," allowing for the modeling of higher-order uncertainties. Type-2 fuzzy logic has demonstrated superior performance in scenarios where uncertainty is multifaceted and intricate, like handling data with varying levels of reliability or precision [8, 9].

This paper presents a significant advancement in the realm of optimization algorithms by proposing an enhancement to the fuzzy artificial fish swarm algorithm. The integration of shadowed type-2 fuzzy logic showcases the algorithm's improved adaptability and performance [10, 11]. The experimental results underscore the advantages of this enhancement, positioning it as a valuable addition to the toolkit of optimization techniques. This research not only contributes to the field's theoretical foundations but also holds practical implications for solving real-world optimization challenges.

The paper is organized as follows: In the next section, we briefly overview the artificial fish swarm algorithm. In Sect. 3, we describe our algorithm: the shadowed type-2 fuzzy logic in the fuzzy artificial fish swarm algorithm. In Sect. 4, we describe experiments of our algorithm with a variety of benchmark data sets. Conclusion section completes our paper.

2 Artificial Fish Swarm Algorithm

Drawing from Li's 2002 study that initially explored the concept of inspiration in food gathering by simulating a group of individuals on water, Yazdani and his team, in 2012, introduced the Artificial Fish Swarm Algorithm (AFSA) algorithm [12]. This algorithm is primarily designed to locate food based on its inherent properties, with a key emphasis on each member's ability to individually scan for prey within their visual range.

The fundamental idea behind the original AFSA can be visualized in Fig. 1, showing each movement executed by a fish within the algorithm. In this representation, 'X' denotes the current position of an individual, often referred to as an artificial fish (AF). The algorithm's execution involves a step denoted as 'S' and the visual distance referred to as 'Visual (V).' Each AF examines the search space within its visual range.

In this context, the optimization criterion for each movement in a fish is determined by whether a new position is superior to its current location. If the new position proves better, the fish will make the corresponding move.

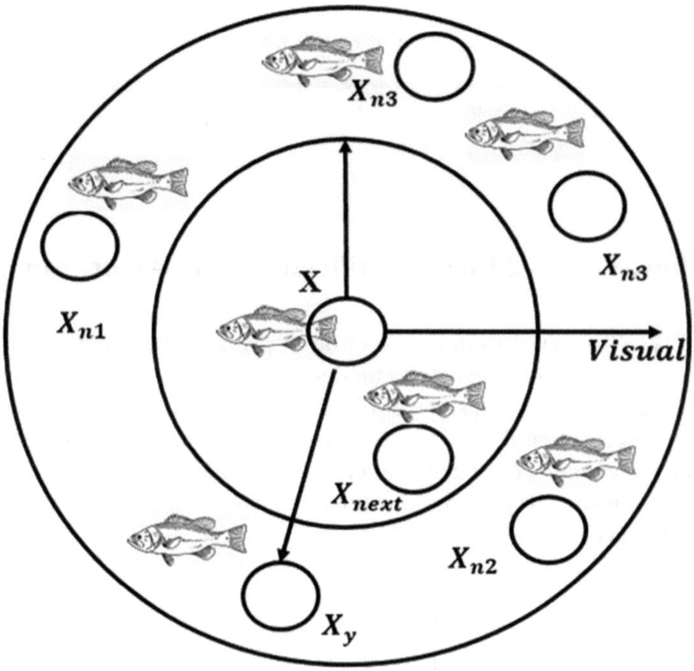

Fig. 1 Visual depiction of the motion patterns of a synthetic fish

In the AFSA algorithm, 'n' stands for a stochastic distribution in each generation within the search area. When looking for an optimal solution, the search space focuses on areas with a higher density of fish. AFSA exhibits four significant behaviors, namely: *preying, swarming, following, and random* behavior. The dynamics of AFSA are governed by Eqs. (1–7).

$$X_j = X_i^{(t)} + V \times rand() \tag{1}$$

$$X_i^{(t+1)} = X_i^{(t)} + S \times rand().* \frac{X_j - X_i^{(t)}}{||X_j - X_i^{(t)}||} \tag{2}$$

$$X_i^{(t+1)} = X_i^{(t)} + V.*rand() \tag{3}$$

$$x_{cd} = \frac{(\sum_{j=1}^{n_f} x_{jd})}{n_f} \tag{4}$$

$$\{Y_c > Y_i \frac{n_f}{n} > \delta \tag{5}$$

$$X_i^{(t+1)} = X_i^{(t)} + S \times rand().* \frac{X_c - X_i^{(t)}}{||X_c - X_i^{(t)}||} \tag{6}$$

$$X_i^{(t+1)} = X_i^{(t)} + V \times rand() \tag{7}$$

3 Shadowed Type-2 Fuzzy Artificial Swarm Algorithm

Shadowed Type 2 fuzzy system is a solution to reduce the computational cost, this work focuses mainly on being able to model GT2-FS using only 2 α-planes, eliminating the excessive precision, and the selection of the optimal α-planes is performed through the concepts that Pedrycz has proposed of the shadowed set theory.

Equation (8) explains the mathematical form of the shaded sets, this formula is built from 2 α-cuts on a Type-1 fuzzy set, with α and β values, based on these α-cuts and three intervals.

$$S_{\mu_A}(x) = \begin{cases} 1, if\ \mu_A(x) \geq \alpha \\ 0, if\ \mu_A(x) \leq \beta \\ [0, 1], if\ \alpha \leq \mu_A(x) \geq \beta \end{cases} \tag{8}$$

These intervals can be interpreted as three regions, the elevated region for the membership degree of 1, the reduced region for the membership degree of 0 and for the shadowed area the membership degree is in [0, 1].

Figure 2 represents the SFAFSA created for experimentation.

The Fig. 3 shows the flowchart of the fish algorithm and the integration of the fuzzy system used to dynamically move the parameters.

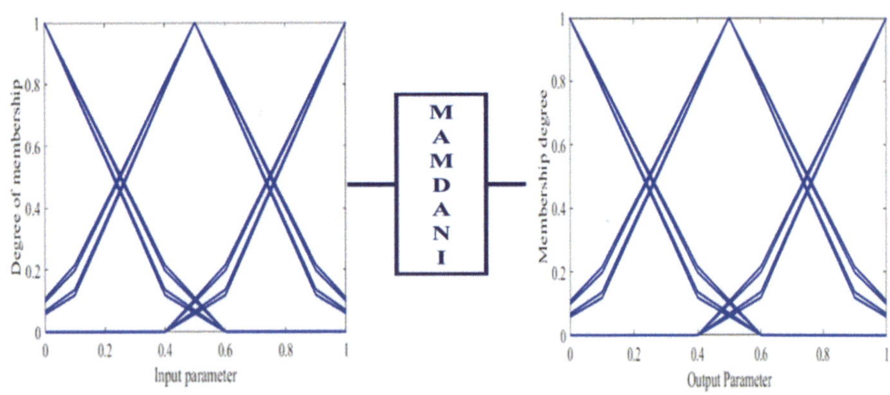

Fig. 2 Fuzzy system SFAFSA

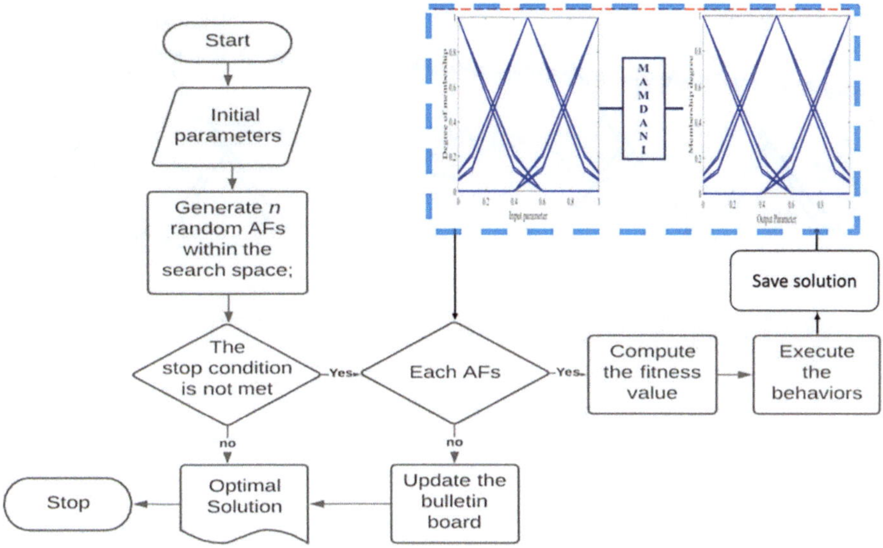

Fig. 3 Flowchart of the procedure in the SFAFSA

The suggested approach for dynamically adjusting the S and V parameters of the AFSA is separated into four SFAFSA components, aiming to assess how the S and V parameters influence the algorithm. The proposal employs triangular Membership Functions (MFs). Figures 4 and 5 depict the visualization of each individual SFAFSA proposed (a) and their corresponding rules (b).

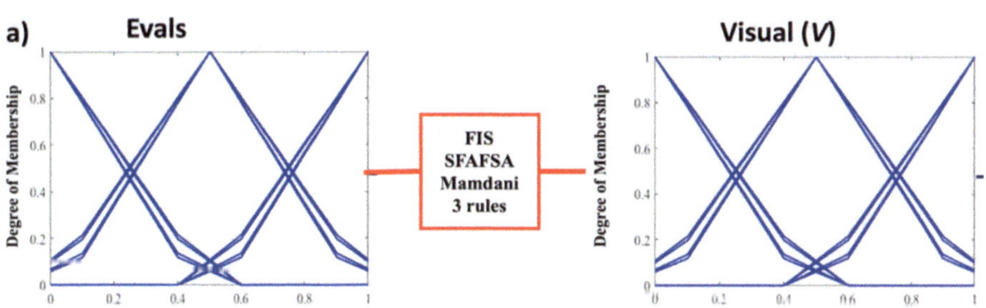

b)

1.- If (Evals is Low) Then (V is High)
2.-If (Evals is Medium) Then (V is Medium)
3.- If (Evals is High) Then (V is Low)

Fig. 4 SFAFSA Adjusting V in a decreasing manner

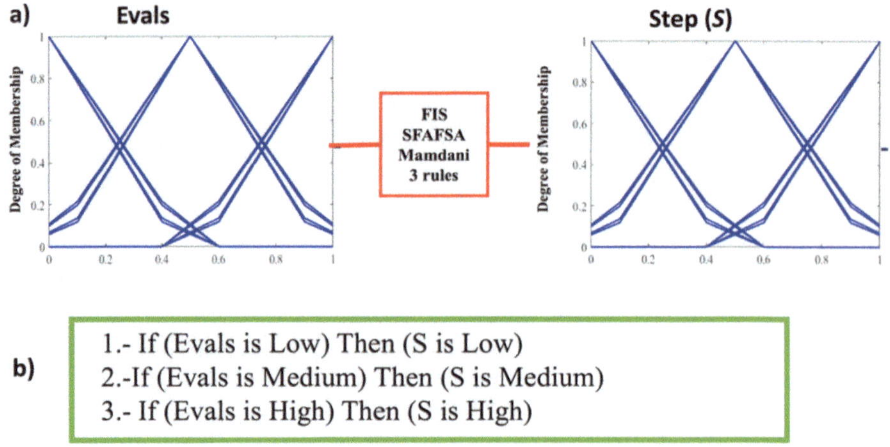

Fig. 5 SFAFSA Gradually changing S

4 Benchmark Sets of Functions and Experimental Results

The focal point of examination in this paper is the set of benchmark functions, and this section outlines a total of six well-established benchmark functions. The primary aim is to assess the performance of SAFSA using these functions, which include Ackley, Giewank, Quarticnoise, Rastrigin, Rosenbrock, and Spherical. All these functions were assessed across dimensions of 10, 30, and 50 (artificial fish). Table 1 presents the equations corresponding to each of these benchmark cases.

The objective of this experimentation is to be able to determine the behavior of each of the FLS when combined with the AFSA. The characteristics used for experimentation are the following: generations are 5000, and dynamic *S* and *V* depending on the T1FLS used.

Table 1 Benchmark sets of functions

f_x	Function	Equation
F1	Ackley	$f(x) = -20e^{-0.2\sqrt{\frac{1}{n_x}\sum_{j=1}^{n_x} x_j^2}} - \frac{1}{e^{n_x}}\sum_{j=1}^{n_x} cos(2\pi x_j) + 20 + e$
F2	Griewank	$f(x) = \sum_{i=1}^{d} \frac{x_i^2}{4000} - \prod_{i=1}^{d} cos\left(\frac{x_i}{\sqrt{i}}\right) + 1$
F3	Quarticnoise	$f(x) = \sum_{i=1}^{d} i x_i^4 + rand[0, 1]$
F4	Rastrigin	$f(x) = 10d + \sum_{i=1}^{d}\left[x_i^2 - 10cos(2\pi x_i)\right]$
F5	Rosenbrock	$f(x) = \sum_{i=1}^{d-1}\left[100(x_{i+1} - x_i^2)^2 + (x_i - 1)^2\right]$
F6	Sphere	$f(x) = \sum_{i=1}^{d} x_i^2$

The aim of this experimentation is to determine whether the adaptation of parameters V and S using the proposed SFAFSA is, on Mean, better than the adaptation of the previously used FAFSA fuzzy systems.

The following tables show a comparison of results obtained in previous work [13].

Table 1 performs comparisons for the variable V using V FAFSA Dec and V SFAFSA Dec, as well as the comparison of the dynamic parameter S using S FAFSA Inc and S SFAFSA Inc, all of this using a dimension of 10.

From the information shown in Table 2 we can see in bold the best error between.

The results obtained for dimensions of 30 can be seen in Table 3, comparisons for the variable V using V FAFSA Dec and V SFAFSA Dec, as well as the comparison of the dynamic parameter S using S FAFSA Inc and S SFAFSA Inc.

Table 2 Outcomes while adjusting the values of "V" and "S" across 10 Dimensions

10 dimensions					
f_x	Performance index	V FAFSA Dec [13]	V SFAFSA Dec	S FAFSA Inc [13]	S SFAFSA Inc
F1	Mean	$2.1 \times 10^{+0}$	$1.3 \times 10^{+0}$	$2.9 \times 10^{+0}$	$2.4 \times 10^{+0}$
	SD	9.1×10^{-1}	7.3×10^{-1}	7.5×10^{-1}	$2.5 \times 10^{+0}$
	Best result	8.2×10^{-4}	7.0×10^{-4}	$7.5 \times 10^{+0}$	$7.2 \times 10^{+0}$
	Worsr result	$3.5 \times 10^{+0}$	$2.8 \times 10^{+0}$	$4.3 \times 10^{+0}$	$5.4 \times 10^{+0}$
F2	Mean	1.6×10^{-1}	3.3×10^{-1}	7.6×10^{-2}	8.5×10^{-2}
	SD	3.3×10^{-1}	5.9×10^{-1}	2.4×10^{-1}	1.4×10^{-1}
	Best result	8.3×10^{-6}	9.8×10^{-5}	4.2×10^{-7}	6.5×10^{-6}
	Worsr result	$1.0 \times 10^{+0}$	$2.1 \times 10^{+0}$	$1.0 \times 10^{+0}$	9.7×10^{-1}
F3	Mean	$1.8 \times 10^{+0}$	$1.9 \times 10^{+0}$	$1.8 \times 10^{+0}$	$2.7 \times 10^{+0}$
	SD	2.4×10^{-1}	1.1×10^{-1}	1.8×10^{-1}	4.7×10^{-1}
	Best result	$1.2 \times 10^{+0}$	2.0×10^{-1}	$1.2 \times 10^{+0}$	$7.3 \times 10^{+0}$
	Worsr result	$2.3 \times 10^{+0}$	$3.9 \times 10^{+0}$	$2.1 \times 10^{+0}$	$4.0 \times 10^{+0}$
F4	Mean	$6.9 \times 10^{+1}$	$8.2 \times 10^{+1}$	$1.3 \times 10^{+2}$	$7.9 \times 10^{+3}$
	SD	$8.1 \times 10^{+1}$	$9.8 \times 10^{+1}$	$7.0 \times 10^{+1}$	$8.8 \times 10^{+1}$
	Best result	8.8×10^{-6}	9.9×10^{-6}	9.1×10^{-4}	3.2×10^{-3}
	Worsr result	$1.7 \times 10^{+2}$	$3.0 \times 10^{+3}$	$1.8 \times 10^{+2}$	$1.4 \times 10^{+2}$
F5	Mean	1.1×10^{-3}	4.4×10^{-3}	6.7×10^{-3}	3.5×10^{-3}
	SD	1.5×10^{-3}	5.8×10^{-3}	1.3×10^{-2}	2.9×10^{-3}
	Best result	6.8×10^{-5}	9.8×10^{-5}	8.0×10^{-5}	8.7×10^{-5}
	Worsr result	8.4×10^{-3}	9.7×10^{-3}	6.1×10^{-2}	8.5×10^{-3}
F6	Mean	9.0×10^{-4}	8.9×10^{-4}	1.2×10^{-3}	3.8×10^{-3}
	SD	5.7×10^{-4}	6.3×10^{-4}	1.0×10^{-3}	1.2×10^{-3}
	Best result	2.4×10^{-4}	1.6×10^{-4}	2.7×10^{-4}	3.1×10^{-6}
	Worsr result	3.7×10^{-3}	5.3×10^{-3}	5.6×10^{-3}	8.0×10^{-3}

Table 3 Outcomes while adjusting the values of "V" and "S" across 30 Dimensions

30 Dimensions					
f_x	Performance index	V FAFSA Dec [13]	V SFAFSA Dec	S FAFSA Inc [13]	S SFAFSA Inc
F1	Mean	$2.3 \times 10^{+0}$	$6.6 \times 10^{+0}$	$2.6 \times 10^{+0}$	$8.1 \times 10^{+1}$
	SD	8.0×10^{-1}	9.3×10^{-1}	8.0×10^{-1}	7.5×10^{-1}
	Best result	8.0×10^{-1}	9.9×10^{-1}	$1.2 \times 10^{+0}$	8.2×10^{-1}
	Worsr result	$5.0 \times 10^{+0}$	$9.3 \times 10^{+0}$	$4.6 \times 10^{+0}$	$9.8 \times 10^{+0}$
F2	Mean	5.3×10^{-4}	7.9×10^{-4}	4.4×10^{-4}	9.9×10^{-4}
	SD	3.1×10^{-4}	2.4×10^{-3}	3.3×10^{-4}	6.9×10^{-4}
	Best result	3.3×10^{-6}	8.0×10^{-5}	1.3×10^{-6}	6.8×10^{-6}
	Worsr result	9.9×10^{-4}	9.8×10^{-4}	9.9×10^{-4}	8.7×10^{-4}
F3	Mean	$9.4 \times 10^{+0}$	$6.0 \times 10^{+1}$	$9.3 \times 10^{+0}$	$8.2 \times 10^{+0}$
	SD	4.3×10^{-1}	3.2×10^{-1}	5.5×10^{-1}	6.5×10^{-1}
	Best result	$8.2 \times 10^{+0}$	$9.0 \times 10^{+0}$	$7.3 \times 10^{+0}$	$8.7 \times 10^{+0}$
	Worsr result	$1.0 \times 10^{+1}$	$3.4 \times 10^{+1}$	$1.0 \times 10^{+1}$	$3.3 \times 10^{+1}$
F4	Mean	1.8×10^{-2}	4.8×10^{-1}	8.5×10^{-2}	9.7×10^{-2}
	SD	4.1×10^{-2}	3.3×10^{-1}	2.2×10^{-1}	7.5×10^{-2}
	Best result	6.2×10^{-4}	2.1×10^{-3}	8.8×10^{-4}	9.4×10^{-3}
	Worsr result	2.2×10^{-1}	1.3×10^{-1}	$1.2 \times 10^{+0}$	3.2×10^{-1}
F5	Mean	9.3×10^{-3}	6.8×10^{-3}	6.0×10^{-2}	7.5×10^{-1}
	SD	1.6×10^{-2}	1.0×10^{-2}	1.0×10^{-1}	4.6×10^{-1}
	Best result	8.2×10^{-5}	6.0×10^{-5}	7.6×10^{-5}	4.4×10^{-5}
	Worsr result	6.1×10^{-2}	6.3×10^{-2}	4.6×10^{-1}	6.5×10^{-2}
F6	Mean	2.3×10^{-3}	2.5×10^{-2}	4.8×10^{-3}	8.5×10^{-3}
	SD	4.0×10^{-3}	4.4×10^{-3}	9.9×10^{-3}	2.0×10^{-2}
	Best result	8.5×10^{-5}	8.9×10^{-5}	8.1×10^{-5}	7.3×10^{-5}
	Worsr result	1.8×10^{-2}	3.9×10^{-2}	4.5×10^{-2}	1.86×10^{-1}

The summary of the results for dimension of 50 are shown in the Table 4, comparisons for the variable V using V FAFSA Dec and V SFAFSA Dec, as well as the comparison of the dynamic parameter S using S FAFSA Inc and S SFAFSA Inc.

5 Conclusions

In this work, two Shadowed Type 2 fuzzy systems were developed, and they were separately used to dynamically adapt each of the V and S parameters in the AFSA algorithm. Experimentation was conducted with six mathematical functions previously employed in a prior study. The reason for choosing these functions was to

Table 4 Outcomes while adjusting the values of "V" and "S" across 50 Dimensions

50 Dimensions					
f_x	Performance index	V FAFSA Dec [13]	V SFAFSA Dec	S FAFSA Inc [13]	S SFAFSA Inc
F1	Mean	$2.1 \times 10^{+0}$	$6.1 \times 10^{+0}$	$2.7 \times 10^{+0}$	$6.3 \times 10^{+0}$
	SD	4.8×10^{-1}	8.8×10^{-2}	6.6×10^{-1}	8.1×10^{-1}
	Best result	9.9×10^{-1}	9.9×10^{-1}	$1.8 \times 10^{+0}$	$8.8 \times 10^{+0}$
	Worsr result	$2.9 \times 10^{+0}$	$6.3 \times 10^{+0}$	$4.8 \times 10^{+0}$	$7.3 \times 10^{+0}$
F2	Mean	4.6×10^{-4}	5.7×10^{-4}	6.2×10^{-4}	8.4×10^{-4}
	SD	6.6×10^{-4}	8.9×10^{-4}	3.4×10^{-4}	4.1×10^{-4}
	Best result	-2.6×10^{-3}	-5.6×10^{-3}	1.8×10^{-7}	7.1×10^{-6}
	Worsr result	9.9×10^{-4}	2.3×10^{-3}	9.9×10^{-4}	8.8×10^{-4}
F3	Mean	$1.7 \times 10^{+1}$	$1.0 \times 10^{+1}$	$1.7 \times 10^{+1}$	$1.1 \times 10^{+1}$
	SD	6.1×10^{-1}	2.1×10^{-1}	6.4×10^{-1}	$6.7 \times 10^{+1}$
	Best result	$1.5 \times 10^{+1}$	$1.0 \times 10^{+1}$	$1.6 \times 10^{+1}$	$1.0 \times 10^{+1}$
	Worsr result	$1.8 \times 10^{+1}$	$1.6 \times 10^{+1}$	$1.8 \times 10^{+1}$	$1.6 \times 10^{+1}$
F4	Mean	8.0×10^{-2}	6.1×10^{-2}	6.9×10^{-2}	2.9×10^{-2}
	SD	1.4×10^{-1}	1.4×10^{-1}	8.2×10^{-2}	1.3×10^{-1}
	Best result	4.9×10^{-4}	4.9×10^{-4}	9.5×10^{-4}	4.9×10^{-4}
	Worsr result	5.7×10^{-1}	4.7×10^{-1}	3.1×10^{-1}	$1.9 \times 10^{+1}$
F5	Mean	2.4×10^{-2}	3.5×10^{-2}	1.1×10^{-1}	4.6×10^{-1}
	SD	6.4×10^{-2}	7.4×10^{-2}	2.0×10^{-1}	5.3×10^{-1}
	Best result	2.6×10^{-5}	5.6×10^{-5}	5.9×10^{-6}	9.6×10^{-5}
	Worsr result	2.7×10^{-1}	5.7×10^{-1}	7.4×10^{-1}	4.7×10^{-1}
F6	Mean	3.5×10^{-3}	6.5×10^{-3}	1.5×10^{-2}	6.6×10^{-2}
	SD	6.0×10^{-3}	7.3×10^{-3}	3.3×10^{-2}	7.1×10^{-2}
	Best result	-6.1×10^{-3}	$-5.1 \times 10^{+2}$	1.5×10^{-4}	3.7×10^{-2}
	Worsr result	2.2×10^{-2}	3.2×10^{-1}	1.5×10^{-1}	4.4×10^{-1}

enable a comparison between Type 1 fuzzy systems for dynamically adjusting the V and S parameters. For this experimentation, the fuzzy rules that yielded the Best result results for each parameter were used to make comparisons between the two types of fuzzy systems and for each of the different dimensions.

The experimentation conducted focused on obtaining the mean, SD, Best result, and worst error. Based on the Best result error results, we can observe that for a dimension of 10, the parameter V has three functions in which V SFAFSA Dec outperforms V FAFSA Dec. Regarding the parameter S, it is evident that S FAFSA Inc yields better results in 4 functions compared to the proposed S SFAFSA Inc.

For dimensions of 30, V FAFSA Dec performs better in 4 functions compared to V SFAFSA Dec, and concerning the parameter S, the S FAFSA Inc system outperforms S SFAFSA Inc in 5 functions.

Finally, for 50 dimensions, dynamically adjusting the parameter V results in the V FAFSA Dec system performing better in 4 functions compared to V SFAFSA Dec. The same holds true for the parameter S, with the S FAFSA Inc system outperforming S SFAFSA Inc in 4 functions, while the latter has two functions with the Best result results.

In conclusion, we can say that the proposal provided did not yield the results we were hoping for in terms of improvements in errors across different dimensions and the set of functions used. As a future work, it may be worthwhile to experiment with a combination of different rules or create a two-input fuzzy system to see if it can lead to improved results. In addition, we can try the same approach with other metaheuristics, like [14–17]. Finally, we could try other application areas, such as in [18–23].

References

1. L.A. Zadeh, Fuzzy sets. Inf. Control. **8**, 338–353 (1965)
2. L. Cervantes, C. Caraveo, O. Castillo, Performance comparison between type-1 and type-2 fuzzy logic control applied to insulin pump injection in real time for patients with diabetes. Mathematics **11**(3), 730 (2023)
3. L. Cherroun, M. Nadour, A. Kouzou, M. Boumehraz, *Type-1 and Type-2 Fuzzy Techniques: Application to Robotic Systems*. In Advances in Robust Control and Applications (Singapore, Springer Nature Singapore, 2023), pp. 319–344
4. A. Khemis, T. Boutabba, S. Drid, Model reference adaptive system speed estimator based on type-1 and type-2 fuzzy logic sensorless control of electrical vehicle with electrical differential. Electr. Eng. Electromech. **4**, 19–25 (2023)
5. L.C. Voumik, R. Karthik, A. Ramamoorthy, A. Dutta, A Study on Mathematics Modeling using Fuzzy Logic and Artificial Neural Network for Medical Decision Making System. In 2023 International Conference on Computational Intelligence and Sustainable Engineering Solutions (CISES) (IEEE, 2023), pp. 492–498
6. Y. Poma, P. Melin, *A New Selection and Class Prediction Using Type-1 Fuzzy Logic Applied to a Convolutional Neural Network*. In Hybrid Intelligent Systems Based on Extensions of Fuzzy Logic, Neural Networks and Metaheuristics. (Springer Nature Switzerland, Cham, 2023), pp. 253–261
7. R. Ranjan, V. Avasthi, An expert system based on type-1 fuzzy logic and digital image processing for knowledge based edge and contour detection. Int. J. Eng. **36**(7), 1278–1286 (2023)
8. J.M. Mendel, F. Liu, D. Zhai, α-plane representation for type-2 fuzzy sets: theory and applications. IEEE Trans. Fuzzy Syst. **17**, 1189–1207 (2009). Y. Liu, X. Feng, Y. Yang, Z. Ruan, L. Zhang, K. Li, Solving urban electric transit network problem by integrating Pareto artificial fish swarm algorithm and genetic algorithm. J. Intell. Transp. Syst. **26**(3), 253–268 (2022)
9. J.M. Mendel, R.I. John, F. Liu, Interval type-2 fuzzy logic systems made simple. IEEE Trans. Fuzzy Syst. **14**, 808–821 (2006)
10. W. Pedrycz, From fuzzy sets to shadowed sets: interpretation and computing. Int. J. Intell. Syst. **24**, 48–61 (2009)
11. W. Pedrycz, M. Song, Granular fuzzy models: a study in knowledge management in fuzzy modeling. Int. J. Approx. Reason. **53**, 1061–1079 (2012). L.G. Wang, Q.H. Shi, Parameters analysis of artificial fish swarm algorithm. Comput. Eng. **36**(24), 169–171 (2010)

12. D. Yazdani, M.R. Akbarzadeh-Totonchi, B. Nasiri, M.R. Meybodi, A new artificial fish swarm algorithm for dynamic optimization problems. In 2012 IEEE Congress on Evolutionary Computation (IEEE, 2012), pp. 1–8
13. L. Amador-Angulo, P. Ochoa, C. Peraza, O. Castillo, Fuzzy Dynamic Adaptation of an Artificial Fish Swarm Algorithm for the Optimization of Benchmark Functions. In *Hybrid Intelligent Systems Based on Extensions of Fuzzy Logic, Neural Networks and Metaheuristics* (Springer Nature Switzerland, Cham, 2023), pp. 99–114
14. O. Castillo, E. Lizzarraga, J. Soria, P. Melin, F. Valdez, New approach using ant colony optimization with ant set partition for fuzzy control design applied to the ball and beam system. Inf. Sci. **294**, 203–215 (2015)
15. L. Amador-Angulo, O. Mendoza, J.R. Castro, A. Rodriguez-Diaz, P. Melin, O. Castillo, Fuzzy sets in dynamic adaptation of parameters of a bee colony optimization for controlling the trajectory of an autonomous mobile robot. Sensors **16**(9), 1458 (2016)
16. F. Valdez, J.C. Vazquez, P. Melin, O. Castillo, Comparative study of the use of fuzzy logic in improving particle swarm optimization variants for mathematical functions using co-evolution. Appl. Soft Comput. **52**, 1070–1083 (2017)
17. D. Sanchez, P. Melin, O. Castillo, A grey wolf optimizer for modular granular neural networks for human recognition. Comput. Intel. Neurosci. 2017.https://doi.org/10.1155/2017/4180510
18. K. Tai, A.-R. El-Sayed, M. Biglarbegian, C.I. Gonzalez, O. Castillo, S. Mahmud, Review of recent type-2 fuzzy controller applications. Algorithms **9**(2), 39 (2016)
19. E. Ontiveros, P. Melin, O. Castillo, Comparative study of interval type-2 and general type-2 fuzzy systems in medical diagnosis. Inf. Sci. **525**, 37–53 (2020)
20. J.E. Moreno, M.A. Sanchez, O. Mendoza, A. Rodriguez-Diaz, O. Castillo, P. Melin, J.R. Castro, Design of an interval type-2 fuzzy model with justifiable uncertainty. Inf. Sci. **513**, 206–221 (2020)
21. P. Melin, O. Castillo, Intelligent control of a stepping motor drive using an adaptive neuro-fuzzy inference system. Inf. Sci. **170**(2–4), 133–151 (2005)
22. P. Melin, O. Castillo, A new method for adaptive control of non-linear plants using type-2 fuzzy logic and neural networks. Int. J. Gen Syst **33**(2–3), 289–304 (2004)
23. P. Melin, O. Castillo, Adaptive intelligent control of aircraft systems with a hybrid approach combining neural networks, fuzzy logic and fractal theory. Appl. Soft Comput. **3**(4), 353–362 (2003)

α-Interval Valued Fuzzy Sets and α-Interval Valued Fuzzy Subgroups

Arif Bal, Gökhan Çuvalcıoğlu, and Feride Tuğrul

1 Introduction

In the fight against uncertainty, situations have occurred where binary logic cannot provide an answer. For these reasons, the concept of fuzzy set was introduced and has become the focus of attention of many researchers. While studies on fuzzy sets continue, researchers have encountered some problems. One of these is not being able to precisely determine the degree of membership. Since the exact membership degree of an element cannot be determined in fuzzy set theory, the membership degree of an element is expanded to closed sub-intervals of the unit interval with the help of interval valued fuzzy sets.

In the paper, interval valued fuzzy sets are shortly shown by IVFSs. IVFSs which is one of the generalization of fuzzy sets, were introduced in [1–4]. Many authors studied on the properties of IVFSs [5–10]. The examination of relation between IVFSs and different structures is crucial. Particularly, the basic topological features of IVFSs were researched in [11].

Rosenfeld introduced the definition of fuzzy subgroups in [12]. He gave the relations between classical groups and fuzzy groups by using level subsets in [12]. The concept of IVF subgroups was identified in [13]. IVF subgroups, their algebraic

A. Bal (✉)
Department of Motor Vehicles and Transportation Technologies, Vocational School of Technical Sciences, Mersin University, Mersin, Turkey
e-mail: arif.bal.math@gmail.com

G. Çuvalcıoğlu
Faculty of Science, Department of Mathematics, Mersin University, Mersin, Turkey

F. Tuğrul
Faculty of Science, Department of Mathematics, Kahramanmaraş Sütçü İmam University, Kahramanmaraş, Turkey

P. Melin and O. Castillo (eds.), *New Directions on Hybrid Intelligent Systems Based on Neural Networks, Fuzzy Logic, and Optimization Algorithms*, Studies in Computational Intelligence 1146, https://doi.org/10.1007/978-3-031-53713-4_4

properties and fuzzy sets were studied by many authors [14–16]. D(I) indicates that all closed sub-intervals of unit interval $I = [0, 1]$.

2 Preliminaries

Definition 1: [14] $S, T \in D(I)$,

i. $S \leq T :\Leftrightarrow S^L \leq T^L \text{and} S^U \leq T^U$

$$S \leq T \not\Leftrightarrow S \subseteq T$$

Definition 2: [14] I is an index set.

For $i \in I$, $S, T, S_i \in D(I)$,

i. $\text{rmax}(S, T) = \left[\max\left(S^L, T^L\right), \max(S^U, T^U)\right]$
ii. $\text{rmin}(S, T) = \left[\min\left(S^L, T^L\right), \min(S^U, T^U)\right]$
iii. $\text{rsup}_{i\in I} S_i = \left[\sup_{i\in I} S_i^L, \sup_{i\in I} S_i^U\right]$
iv. $\text{rinf}_{i\in I} S_i = \left[\inf_{i\in I} S_i^L, \inf_{i\in I} S_i^U\right]$

Definition 3: [7] X is a universal set.

$A : X \to D(I)$,

$$A = \{\langle w, A(w)\rangle | w \in X\} = \left\{\langle w, \left[A^L(w), A^U(w)\right]\rangle | w \in X\right\}$$

is called IVFSs. $A^L, A^U : X \to [0, 1]$ are fuzzy sets and $\forall w \in X$, $A^L(w) \leq A^U(w)$. A^L is named lower bound and A^U is named upper bound. The family of all IVFSs on X is denoted by IVFS(X).

Example 1: Let $X = \{k, l, m, n\}$.

$$A = \{< k, [0.0, 0.7] >, < l, [0.2, 0.4] >, < m, [0.3, 0.8] >, < n, [0.5, 1.0] >\}$$

is an IVFS.

Definition 4: [7] X is a universal set.

$A, B \in IVFS(X)$,

i. $A \subseteq B :\Leftrightarrow \forall w \in X,\ A^L(w) \leq B^L(w),\ A^U(w) \leq B^U(w)$

ii. $A = B :\Leftrightarrow \forall w \in X, A^L(w) = B^L(w), A^U(w) = B^U(w)$

iii. $A \sqcap B = \{< w, \begin{bmatrix} min\{A^L(w), B^L(w)\}, \\ min\{A^U(w), B^U(w)\} \end{bmatrix} > w \in X\}$

iv. $A \bigsqcup B = \{< w, \begin{bmatrix} max\{A^L(w), B^L(w)\}, \\ max\{A^U(w), B^U(w)\} \end{bmatrix} > w \in X\}$

v. $A^c = \{< w, [1 - A^U(w), 1 - A^L(w)] > w \in X\}$

Definition 5: [7] X is a universal set.

$i \in I \ni$ Iisanindexset,

i. $\sqcap_{i \in I} A_i = \{< w, [\min_{i \in I}\{A_i{}^L(w)\}, \min_{i \in I}\{A_i{}^U(w)\}] > |w \in X\}$

ii. $\bigsqcup_{i \in I} A_i = \{< w, [\max_{i \in I}\{A_i{}^L(w)\}, \max_{i \in I}\{A_i{}^U(w)\}] > |w \in X\}$

The commutative and distributive properties of intersection and union take an important role in set theory. The properties mentioned above are applied for IVFSs in Proposition1.

Proposition 1: [7] X is a universal set.

$i \in I, A_i, B_i \in IVFS(X) \ni$ Iisanindexset and $A, B \in IVFS(X)$,

i. $A \sqcap B = B \sqcap A$

ii. $A \bigsqcup B = B \bigsqcup A$

iii. $A \sqcap (\bigsqcup_{i \in I} B_i) = \bigsqcup_{i \in I}(A \sqcap B_i)$

iv. $A \bigsqcup (\sqcap_{i \in I} B_i) = \sqcap_{i \in I}(A \bigsqcup B_i)$

Definition 6: [15] X is a universal set. $\forall[\eta_1, \eta_2] \in [0, 1]$,

$$A_{[\eta_1, \eta_2]} = \{w \in X | A^L(w) \leq \eta_1 \text{and} A^U(w) \leq \eta_2\}$$

$A_{[\eta_1, \eta_2]}$ is called $[\eta_1, \eta_2]$-level subset of A.

The definition of $A^L_{\eta_1}, A^U_{\eta_2}, B^L_{\eta_1}, B^U_{\eta_2}$ are given below.

i. $A^L_{\eta_1} = \{w \in X | A^L(w) \leq \eta_1\}$

ii. $A^U_{\eta_2} = \{w \in X | A^U(w) \leq \eta_2\}$

iii. $B^L_{\eta_1} = \{w \in X | B^L(w) \leq \eta_1\}$

iv. $B^U_{\eta_2} = \{w \in X | B^U(w) \leq \eta_2\}$

Proposition about $[\eta_1, \eta_2]$-level subset of A is indicated below.

Proposition 2: [15] X is a universal set.

$A, B \in IVFS(X)$ and $[\eta_1, \eta_2] \in [0, 1]$,

i. $w \in A_{[\eta_1, \eta_2]} \iff A(w) \leq [\eta_1, \eta_2]$

ii. $A \bigsqcup B_{[\eta_1, \eta_2]} = A_{[\eta_1, \eta_2]} \cap B_{[\eta_1, \eta_2]}$

iii. $A \sqcap B_{[\eta_1, \eta_2]} = A_{[\eta_1, \eta_2]} \cup B_{[\eta_1, \eta_2]} \cup (A^L_{\eta_1} \cap B^U_{\eta_2}) \cup (A^U_{\eta_2} \cap B^L_{\eta_1})$

3 α-Interval Valued Fuzzy Sets

The features of some special elements of D(I) are examined. D(I) has not different properties of the family of closed sub-interval of I. The categorization of the elements of D(I) is difficult. The definition of $D(I_\alpha)$ is introduced below. $D(I_\alpha)$ represents α-interval valued set that is shown shortly α-IVS.

Definition 7: $D(I_\alpha) = \{[S^L, S^U; \alpha]\alpha \in I\}$ is named α-IVS.

Used impression to make notation clear is given below.

$$\{[S^L, S^U; \alpha]\alpha \in I\} = \{[S; \alpha]S \in D(I) \text{and} \alpha \in S\}$$

After the definition above, order relation on that set is shown like,

Definition 8: $[S; \alpha], [T; \alpha] \in D(I_\alpha)$,

$$[S; \alpha] \leq [T; \alpha] :\Longleftrightarrow S^L \leq T^L \text{and} S^U \geq T^U$$

From definition, it is obvious that

$$[S; \alpha] < [T; \alpha] \Longleftrightarrow S^L < T^L, S^U \geq T^U$$

$$\text{or} S^L \leq T^L, S^U > T^U \text{or} S^L < T^L, S^U > T^U$$

Proposition 3: $(D(I_\alpha), \leq)$ is partial ordered set.

Proof: $[S; \alpha], [T; \alpha], [V; \alpha] \in D(I_\alpha)$,

i. $S^L \leq S^L \text{and} S^U \geq S^U \Rightarrow [S; \alpha] \leq [S; \alpha]$
ii. $[S; \alpha] \leq [T; \alpha] \text{and} [T; \alpha] \leq [S; \alpha]$

$$\Rightarrow S^L \leq T^L, S^U \geq T^U \text{and} T^L \leq S^L, T^U \geq S^U$$

$$\Rightarrow S^L = T^L \text{and} S^U = T^U \Rightarrow [S; \alpha] = [T; \alpha]$$

iii. $[S; \alpha] \leq [T; \alpha] \text{and} [T; \alpha] \leq [V; \alpha]$

$$\Rightarrow S^L \leq T^L, S^U \geq T^U \text{and} T^L \leq V^L, T^U \geq V^U$$

$$\Rightarrow S^L \leq V^L and S^U \geq V^U \Rightarrow [S;\alpha] \leq [V;\alpha]$$

After discussions on order relation, bounds on $D(I_\alpha)$ are introduced below.

Definition 9: $[S;\alpha], [T;\alpha] \in D(I_\alpha)$,

i. $\inf\{[S;\alpha],[T;\alpha]\} = \left[\inf\{S^L, T^L\}, \sup\{S^U, T^U\};\alpha\right]$
ii. $\sup\{[S;\alpha],[T;\alpha]\} = \left[\sup\{S^L, T^L\}, \inf\{S^U, T^U\};\alpha\right]$

Lemma 1: $(D(I_\alpha), \wedge, \vee)$ is complete lattice whose units are $[0, 1;\alpha]$ and $[\alpha, \alpha;\alpha]$.

Proposition 4: $\alpha \in I$,

$$\bigcup_{\alpha\in I} D(I_\alpha) = D(I)$$

Proof: $S \in \bigcup_{\alpha\in I} D(I_\alpha)$,

$$\forall\alpha \in I, D(I_\alpha) \subseteq D(I) \Rightarrow \bigcup_{\alpha\in I} D(I_\alpha) \subseteq D(I)$$

$S \in D(I)$,

$$\alpha = \frac{S^L + S^U}{2} is chosen,$$

$$S^L \leq \alpha \leq S^U \Rightarrow D(I) \subseteq D(I_\alpha) \Rightarrow D(I) \subseteq \bigcup_{\alpha\in I} D(I_\alpha)$$

Remark 1: The intersection and union of the family of α-IVS are again α-IVS.

The functions that satisfy following conditions are named negation function.

Definition 10: L is complete lattice with units 0 and 1. $\mathcal{N} : L \rightarrow L$ and $\forall w, z \in L$,

i. $\mathcal{N}(0) = 1$ and $\mathcal{N}(1) = 0$
ii. $\mathcal{N}(w) \leq \mathcal{N}(z) :\Longleftrightarrow w \geq z$
iii. $\mathcal{N}(\mathcal{N}(w)) = w$

The following function is used to give the definition of negation function on $D(I_\alpha)$, $[S; \alpha] \in D(I_\alpha)$,

$$\mathcal{N}([S; \alpha]) = \left[\alpha - S^L, 1 + \alpha - S^U; \alpha\right]$$

is a function $D(I_\alpha) \to D(I_\alpha)$. Well-defined of $\mathcal{N}$ is pointed below.

i. i.$S^L \leq \alpha \Rightarrow 0 \leq \alpha - S^L \leq \alpha$
ii. Suppose that $1 + \alpha - S^U < \alpha$ thus,

   $1 - S^U < 0 \Rightarrow 1 < S^U$ is contradiction. We get that $1 + \alpha - S^U \geq \alpha$

iii. Suppose that $1 + \alpha - S^U > 1$ thus,

   $\alpha - S^U > 0 \Rightarrow \alpha > S^U$ is contradiction. We get that $1 + \alpha - S^U \leq 1$ consequences from above, we get that;

$$0 \leq \alpha - S^L \leq \alpha \leq 1 + \alpha - S^U \leq 1$$

$$\Rightarrow \left[\alpha - S^L, 1 + \alpha - S^U; \alpha\right] = \mathcal{N}([S; \alpha]) \in D(I_\alpha)$$

it is claimed that $\mathcal{N}$ is a negation function on $D(I_\alpha)$ from discussions from above.

Proposition 5: $[S; \alpha] \in D(I_\alpha)and\mathcal{N} : D(I_\alpha) \to D(I_\alpha)$,

$$\mathcal{N}([S; \alpha]) = \left[\alpha - S^L, 1 + \alpha - S^U; \alpha\right]$$

Proof: $[S; \alpha], [T; \alpha] \in D(I_\alpha)$,

i. $\mathcal{N}([0, 1; \alpha]) = [\alpha - 0, 1 + \alpha - 1; \alpha] = [\alpha, \alpha; \alpha]$

$$\mathcal{N}([\alpha, \alpha; \alpha]) = [\alpha - \alpha, 1 + \alpha - \alpha; \alpha] = [0, 1; \alpha]$$

ii. $\mathcal{N}([S; \alpha]) \leq \mathcal{N}([T; \alpha])$

$$\Leftrightarrow \left[\alpha - S^L, 1 + \alpha - S^U; \alpha\right] \leq \left[\alpha - T^L, 1 + \alpha - T^U; \alpha\right]$$

$$\Leftrightarrow \alpha - S^L \leq \alpha - T^L \text{and} 1 + \alpha - S^U \geq 1 + \alpha - T^U$$

$$\Leftrightarrow S^L \geq T^L \text{and} S^U \leq T^U \Leftrightarrow [S; \alpha] \geq [T; \alpha]$$

iii. $\mathcal{N}(\mathcal{N}([S;\alpha])) = \mathcal{N}\left(\left[\alpha - S^L, 1 + \alpha - S^U; \alpha\right]\right)$

$$= \left[\alpha - \left(\alpha - S^L\right), 1 + \alpha - \left(1 + \alpha - S^U\right); \alpha\right] = \left[S^L, S^U; \alpha\right] = [S;\alpha]$$

it is seen that $\mathcal{N}$ is negation function.

Definition 11: X is a universal set.

$[A;\alpha] : X \to D(I_\alpha)$ is a function.

$$[A;\alpha] = \left\{\left[\langle w, \left[A^L(w), A^U(w)\right]\rangle; \alpha\right] w \in X\right\}$$

Below expression is used to make notation clear,

$$\left\{\left[\langle w, \left[A^L(w), A^U(w)\right]\rangle; \alpha\right] w \in X\right\} = \{[\langle w, A(w)\rangle; \alpha] w \in X\}$$

$[A;\alpha]$ that defined on X is named $\alpha-$ IVFSs.
$\alpha-$ IVFS(X) denotes the family of $\alpha-$ IVFSs on X.
The basic operations on these set are given following,

Definition 12: X is a universal set.

$[A;\alpha], [B;\alpha] \in \alpha-$ IVFS(X). Kisindexset$\forall\eta \in K$,

i. $\left[A^c;\alpha\right] = \left\{\left[< w, \left[\alpha - A^L(w), 1 + \alpha - A^U(w)\right] >; \alpha\right] w \in X\right\}$
ii. $[A;\alpha] \subseteq [B;\alpha] :\Leftrightarrow \forall w \in X, A^L(w) \leq B^L(w) \text{and} A^U(w) \geq B^U(w)$
iii. $[A;\alpha] = [B;\alpha] :\Leftrightarrow \forall w \in X, A^L(w) = B^L(w) \text{and} A^U(w) = B^U(w)$
iv. $[A \sqcap B;\alpha] = \left\{\left[< w, \left[\inf\left\{A^L(w), B^L(w)\right\}, \sup\left\{A^U(w), B^U(w)\right\}\right] >; \alpha\right] w \in X\right\}$
v. $\left[A \bigsqcup B;\alpha\right] = \left\{\left[< w, \left[\sup\left\{A^L(w), B^L(w)\right\}, \inf\left\{A^U(w), B^U(w)\right\}\right] >; \alpha\right] w \in X\right\}$
vi. $\left[\sqcap_{\eta\in K} A_\eta;\alpha\right] = \left\{\left[< w, \left[\bigwedge_{\eta\in K} {A_\eta}^L(w), \bigvee_{\eta\in K} {A_\eta}^U(w)\right] >; \alpha\right] w \in X\right\}$
vii. $\left[\bigsqcup_{\eta\in K} A_\eta;\alpha\right] = \left\{\left[< w, \left[\bigvee_{\eta\in K} {A_\eta}^L(w), \bigwedge_{\eta\in K} {A_\eta}^U(w)\right] >; \alpha\right] w \in X\right\}$

Example 2: Let X = {k, l, m, n}.

$$[A; 0.5] = \left\{\begin{array}{c} [< k, [0.2, 0.7] >; 0.5], [< l, [0.1, 0.8] >; 0.5], \\ [< m, [0.45, 0.6] >; 0.5], [< n, [0.35, 1] >; 0.5] \end{array}\right\}$$

$$[B; 0.5] = \left\{ \begin{array}{l} [< k, [0.0, 0.9] >; 0.5], [< l, [0.25, 0.55] >; 0.5], \\ [< m, [0.5, 0.9] >; 0.5], [< n, [0.3, 0.65] >; 0.5] \end{array} \right\}$$

[A; 0.5]and[B; 0.5] are 0.5-IVFSs;

i. $\left[A^c; 0.5\right] = \left\{ \begin{array}{l} [< k, [0.3, 0.8] >; 0.5], [< l, [0.4, 0.7] >; 0.5], \\ [< m, [0.05, 0.9] >; 0.5], [< n, [0.15, 0.5] >; 0.5] \end{array} \right\}$

ii. $[A \sqcap B; 0.5] = \left\{ \begin{array}{l} [< k, [0.0, 0.9] >; 0.5], [< l, [0.1, 0.8] >; 0.5], \\ [< m, [0.45, 0.9] >; 0.5], [< n, [0.3, 1] >; 0.5] \end{array} \right\}$

iii. $\left[A \bigsqcup B; 0.5\right] = \left\{ \begin{array}{l} [< k, [0.2, 0.7] >; 0.5], [< l, [0.25, 0.55] >; 0.5], \\ [< m, [0.5, 0.6] >; 0.5], [< n, [0.35, 0.65] >; 0.5] \end{array} \right\}$

Proposition 6: X is a universal set.

$[A; \alpha], [B; \alpha], [C; \alpha] \in \alpha-$ IVFS(X) and Kisindexset$\eta \in$ K,

i. $[A \sqcap B; \alpha] = [B \sqcap A; \alpha]$
ii. $\left[A \bigsqcup B; \alpha\right] = \left[B \bigsqcup A; \alpha\right]$
iii. $[A; \alpha] \sqcap \left(\left[B \bigsqcup C; \alpha\right]\right) = ([A \sqcap B; \alpha]) \bigsqcup ([A \sqcap C; \alpha])$
iv. $[A; \alpha] \bigsqcup ([B \sqcap C; \alpha]) = \left(\left[A \bigsqcup B; \alpha\right]\right) \sqcap \left(\left[A \bigsqcup C; \alpha\right]\right)$
v. $[A; \alpha] \sqcap \left(\left[\bigsqcup_{\eta \in K} B_\eta; \alpha\right]\right) = \left[\bigsqcup_{\eta \in K} (A \sqcap B_\eta); \alpha\right]$
vi. $[A; \alpha] \bigsqcup \left(\left[\sqcap_{\eta \in K} B_\eta; \alpha\right]\right) = \left[\sqcap_{\eta \in K} \left(A \bigsqcup B_\eta\right); \alpha\right]$

Proof: $[A; \alpha], [B; \alpha], [C; \alpha] \in \alpha-$IVFS(X),

i. $[A \sqcap B; \alpha] = \left\{\left[< w, \left[\inf\left\{A^L(w), B^L(w)\right\}, \sup\left\{A^U(w), B^U(w)\right\}\right] >; \alpha\right] w \in X\right\}$

$$= \left\{\left[< w, \left[\inf\left\{B^L(w), A^L(w)\right\}, \sup\left\{B^U(w), A^U(w)\right\}\right] >; \alpha\right] w \in X\right\} = [B \sqcap A; \alpha]$$

ii. $\left[A \bigsqcup B; \alpha\right]$
$= \left\{\left[< w, \left[\sup\left\{A^L(w), B^L(w)\right\}, \inf\left\{A^U(w), B^U(w)\right\}\right] >; \alpha\right] w \in X\right\}$

$$= \left\{\left[< w, \left[\sup\left\{B^L(w), A^L(w)\right\}, \inf\left\{B^U(w), A^U(w)\right\}\right] >; \alpha\right] w \in X\right\} = \left[B \bigsqcup A; \alpha\right]$$

iii. $[A; \alpha] \sqcap \left(\left[B \bigsqcup C; \alpha\right]\right)$

$$= \left\{\left[< w, \left[\begin{array}{l} \inf\left\{A^L(w), \sup\left\{B^L(w), C^L(w)\right\}\right\}, \\ \sup\left\{A^U(w), \inf\left\{B^U(w), C^U(w)\right\}\right\} \end{array}\right] >; \alpha\right] w \in X\right\}$$

$$= \left\{\left[< w, \begin{bmatrix} \sup\{\inf\{A^L(w), B^L(w)\}, \inf\{A^L(w), C^L(w)\}\}, \\ \inf\{\sup\{A^U(w), B^U(w)\}, \sup\{A^U(w), C^U(w)\}\} \end{bmatrix} >; \alpha\right] w \in X\right\}$$

$$= \left\{\left[< w, \begin{bmatrix} \inf\{A^L(w), B^L(w)\}, \\ \sup\{A^U(w), B^U(w)\} \end{bmatrix} >; \alpha\right] w \in X\right\} \bigsqcup$$
$$\left\{\left[< w, \begin{bmatrix} \inf\{A^L(w), C^L(w)\}, \\ \sup\{A^U(w), C^U(w)\} \end{bmatrix} >; \alpha\right] w \in X\right\}$$

$$([A \sqcap B; \alpha]) \bigsqcup ([A \sqcap C; \alpha])$$

iv. $[A; \alpha] \bigsqcup ([B \sqcap C; \alpha])$

$$= \left\{\left[< w, \begin{bmatrix} \sup\{A^L(w), \inf\{B^L(w), C^L(w)\}\}, \\ \inf\{A^U(w), \sup\{B^U(w), C^U(w)\}\} \end{bmatrix} >; \alpha\right] w \in X\right\}$$

$$= \left\{\left[< w, \begin{bmatrix} \inf\{\sup\{A^L(w), B^L(w)\}, \sup\{A^L(w), C^L(w)\}\}, \\ \sup\{\inf\{A^U(w), B^U(w)\}, \inf\{A^U(w), C^U(w)\}\} \end{bmatrix} >; \alpha\right] w \in X\right\}$$

$$= \left\{\left[< w, \begin{bmatrix} \sup\{A^L(w), B^L(w)\}, \\ \inf\{A^U(w), B^U(w)\} \end{bmatrix} >; \alpha\right] w \in X\right\} \sqcap$$
$$\left\{\left[< w, \begin{bmatrix} \sup\{A^L(w), C^L(w)\}, \\ \inf\{A^U(w), C^U(w)\} \end{bmatrix} >; \alpha\right] w \in X\right\}$$
$$= \left(\left[A \bigsqcup B; \alpha\right]\right) \sqcap \left(\left[A \bigsqcup C; \alpha\right]\right)$$

v. $[A; \alpha] \sqcap \left(\left[\bigsqcup_{\eta\in K} B_\eta; \alpha\right]\right)$

$$= \left\{\left[\langle w, \begin{bmatrix} A^L(w), \\ A^U(w) \end{bmatrix} \rangle; \alpha\right] w \in X\right\} \sqcap \left(\left\{\left[< w, \left[\bigvee_{\eta\in K} B_\eta{}^L(w), \bigwedge_{\eta\in K} B_\eta{}^U(w)\right] >; \alpha\right] w \in X\right\}\right)$$

$$= \left\{\left[\langle w, \left[A^L(w) \wedge \bigvee_{\eta\in K} B_\eta{}^L(w), A^U(w) \vee \bigwedge_{\eta\in K} B_\eta{}^U(w)\right] \rangle; \alpha\right] w \in X\right\}$$

$$= \left\{\left[\langle w, \left[\bigvee_{\eta\in K} A^L(w) \wedge B_\eta{}^L(w), \bigwedge_{\eta\in K} A^U(w) \vee B_\eta{}^U(w)\right] \rangle; \alpha\right] w \in X\right\}$$

$$= \left[\bigsqcup_{\eta\in K}(A \sqcap B_\eta); \alpha\right]$$

vi. $[A; \alpha] \bigsqcup \left(\left[\sqcap_{\eta\in K} B_\eta; \alpha\right]\right)$

$$= \left\{\left[\langle w, \begin{bmatrix} A^L(w), \\ A^U(w) \end{bmatrix}\rangle; \alpha\right] w \in X\right\} \bigsqcup$$

$$\left(\left\{\left[< w, \left[\bigwedge_{\eta\in K} B_\eta{}^L(w), \bigvee_{\eta\in K} B_\eta{}^U(w)\right] >; \alpha\right] w \in X\right\}\right)$$

$$= \left\{\left[\langle w, \left[A^L(w) \vee \bigwedge_{\eta\in K} B_\eta{}^L(w), A^U(w) \wedge \bigvee_{\eta\in K} B_\eta{}^U(w)\right]\rangle; \alpha\right] w \in X\right\}$$

$$= \left\{\left[\langle w, \left[\bigwedge_{\eta\in K} A^L(w) \vee B_\eta{}^L(w), \bigvee_{\eta\in K} A^U(w) \wedge B_\eta{}^U(w)\right]\rangle; \alpha\right] w \in X\right\}$$

$$= \left[\sqcap_{\eta\in K}\left(A \bigsqcup B_\eta\right); \alpha\right]$$

Proposition 7: X is a universal set.

$[A; \alpha], [B; \alpha] \in \alpha-$ IVFS(X). Kisindexset$\forall\eta \in K$,

i. $\left[\left(\left[A^c; \alpha\right]\right)^c; \alpha\right] = [A; \alpha]$
ii. $([A \sqcap B; \alpha])^c = \left[A^c \bigsqcup B^c; \alpha\right]$
iii. $\left(\left[A \bigsqcup B; \alpha\right]\right)^c = \left[A^c \sqcap B^c; \alpha\right]$
iv. $\left(\left[\sqcap_{\eta\in K} A_\eta; \alpha\right]\right)^c = \left[\bigsqcup_{\eta\in K} A_\eta^c; \alpha\right]$
v. $\left(\left[\bigsqcup_{\eta\in K} A_\eta; \alpha\right]\right)^c = \left[\sqcap_{\eta\in K} A_\eta^c; \alpha\right]$

Proof: $[A; \alpha], [B; \alpha] \in \alpha-$IVFS(X),

i. $\left[\left(\left[A^c; \alpha\right]\right)^c; \alpha\right]$

$$= \left\{\left[\langle w, \left[\alpha - \left(\alpha - A^L(w)\right), 1 + \alpha - \left(1 + \alpha - A^U(w)\right)\right]\rangle; \alpha\right] w \in X\right\}$$

$$\Rightarrow \left[\left(\left[A^c; \alpha\right]\right)^c; \alpha\right] = \left\{\left[\langle w, \left[A^L(w), A^U(w)\right]\rangle; \alpha\right] w \in X\right\} = [A; \alpha]$$

ii. $$([A \sqcap B; \alpha])^c = \left\{\left[\langle w, \begin{bmatrix} \alpha - \inf\left\{A^L(w), B^L(w)\right\}, \\ 1 + \alpha - \sup\left\{A^U(w), B^U(w)\right\} \end{bmatrix}\rangle; \alpha\right] w \in X\right\}$$

$$= \left\{ \left[\langle w, \begin{bmatrix} \sup\{\alpha - A^L(w), \alpha - B^L(w)\}, \\ \inf\{1 + \alpha - A^U(w), 1 + \alpha - B^U(w)\} \end{bmatrix} \rangle; \alpha \right] w \in X \right\}$$

$$= \left[A^c \bigsqcup B^c; \alpha \right]$$

iii. $$\left([A \bigsqcup B; \alpha] \right)^c = \left\{ \left[\langle w, \begin{bmatrix} \alpha - \sup\{A^L(w), B^L(w)\}, \\ 1 + \alpha - \inf\{A^U(w), B^U(w)\} \end{bmatrix} \rangle; \alpha \right] w \in X \right\}$$

$$= \left\{ \left[\langle w, \begin{bmatrix} \inf\{\alpha - A^L(w), \alpha - B^L(w)\}, \\ \sup\{1 + \alpha - A^U(w), 1 + \alpha - B^U(w)\} \end{bmatrix} \rangle; \alpha \right] w \in X \right\}$$

$$= \left[A^c \sqcap B^c; \alpha \right]$$

iv. $\left([\sqcap_{\eta \in K} A_\eta; \alpha] \right)^c$

$$= \left\{ \left[\langle w, \left[\alpha - \bigwedge_{\eta \in K} {A_\eta}^L(w), 1 + \alpha - \bigvee_{\eta \in K} {A_\eta}^U(w) \right] \rangle; \alpha \right] w \in X \right\}$$

$$= \left\{ \left[\langle w, \left[\bigvee_{\eta \in K} \alpha - {A_\eta}^L(w), \bigwedge_{\eta \in K} 1 + \alpha - {A_\eta}^U(w) \right] \rangle; \alpha \right] w \in X \right\} = \left[\bigsqcup_{\eta \in K} A_\eta^c; \alpha \right]$$

v. $\left(\left[\bigsqcup_{\eta \in K} A_\eta; \alpha \right] \right)^c$

$$= \left\{ \left[\langle w, \left[\alpha - \bigvee_{\eta \in K} {A_\eta}^L(w), 1 + \alpha - \bigwedge_{\eta \in K} {A_\eta}^U(w) \right] \rangle; \alpha \right] w \in X \right\}$$

$$= \left\{ \left[\langle w, \left[\bigwedge_{\eta \in K} \alpha - {A_\eta}^L(w), \bigvee_{\eta \in K} 1 + \alpha - {A_\eta}^U(w) \right] \rangle; \alpha \right] w \in X \right\} = \left[\sqcap_{\eta \in K} A_\eta^c; \alpha \right]$$

Proposition 8: X is a universal set.

$0_X : X \to [0, 1; \alpha]and1_X : X \to [\alpha, \alpha; \alpha]$ are fixed functions.

i. $(0_X)^c = 1_X$
ii. $(1_X)^c = 0_X$

Proof: It is clear.

Definition 13: X is a universal set. $[A;\alpha] \in \alpha-$ IVFS(X).

$[A;\alpha]$ has sup-property

$$:\Longleftrightarrow \forall w \in X, \exists[\eta_1,\eta_2;\alpha] \in D(I_\alpha) \ni [A(w);\alpha] = [\eta_1,\eta_2;\alpha]$$

Definition 14: X is a universal set.

$[A;\alpha] \in \alpha-$ IVFS(X) and $\forall[\eta_1,\eta_2;\alpha] \in D(I_\alpha)$,

$$[A;\alpha]_{[\eta_1,\eta_2;\alpha]} = \left\{w \in XA^L(w) \geq \eta_1 \text{and} A^U(w) \leq \eta_2\right\}$$

$[A;\alpha]_{[\eta_1,\eta_2;\alpha]}$ is named $[\eta_1,\eta_2;\alpha]$-level subset of $[A;\alpha]$. It is clear that $[\eta_1,\eta_2;\alpha]$-level subsets of $[A;\alpha]$ are crisp sets.

Definition 15: X is a universal set.

$[A;\alpha] \in \alpha-$ IVFS(X) and $\forall[\eta_1,\eta_2;\alpha] \in D(I_\alpha)$,
$[A;\alpha]_{[\eta_1,\eta_2;\alpha]}$-level subsets of $[A;\alpha]$,

i. $A^L_{\eta_1} = \left\{w \in XA^L(w) \geq \eta_1\right\}$
ii. $A^U_{\eta_2} = \left\{w \in XA^U(w) \leq \eta_2\right\}$
iii. $B^L_{\eta_1} = \left\{w \in XB^L(w) \geq \eta_1\right\}$
iv. $B^U_{\eta_2} = \left\{w \in XB^U(w) \leq \eta_2\right\}$

Proposition 9: X is a universal set.

$[A;\alpha], [B;\alpha] \in \alpha-$ IVFS(X). $[\eta_1,\eta_2;\alpha] \in D(I_\alpha)$ and I is index set and $i, j \in I$, $\left[\eta_i,\eta_j;\alpha\right] \in D(I_\alpha)$,

i. $w \in [A;\alpha]_{[\eta_1,\eta_2;\alpha]} \Leftrightarrow [A(w);\alpha] \geq [\eta_1,\eta_2;\alpha]$
ii. $[A;\alpha]_{[\eta_1,\eta_2;\alpha]} = A^L_{\eta_1} \cap A^U_{\eta_2}$
iii. $\left(\left[A \bigsqcup B;\alpha\right]\right)_{[\eta_1,\eta_2;\alpha]}$

$$= [A;\alpha]_{[\eta_1,\eta_2;\alpha]} \cup [B;\alpha]_{[\eta_1,\eta_2;\alpha]} \cup \left(A^L_{\eta_1} \cap B^U_{\eta_2}\right) \cup \left(B^L_{\eta_1} \cap A^U_{\eta_2}\right)$$

iv. $([A \sqcap B;\alpha])_{[\eta_1,\eta_2;\alpha]} = [A;\alpha]_{[\eta_1,\eta_2;\alpha]} \cap [B;\alpha]_{[\eta_1,\eta_2;\alpha]}$
v. $A^L_{\eta_1} \supseteq A^L_{\eta_2}$
vi. $A^U_{\eta_1} \subseteq A^U_{\eta_2}$
vii. $\bigcap_{i\in I} A^L{}_{\eta_i} = A^L{}_{\bigwedge_{i\in I}\eta_i}$
viii. $\bigcup_{j\in I} A^U{}_{\eta_j} = A^U{}_{\bigvee_{j\in I}\eta_j}$

Proof:

i. $w \in [A;\alpha]_{[\eta_1,\eta_2;\alpha]} \Leftrightarrow A^L(w) \geq \eta_1 \text{and} A^U(w) \leq \eta_2$

$$\Leftrightarrow \left[A^L(w), A^U(w);\alpha\right] \geq [\eta_1, \eta_2;\alpha] \Leftrightarrow [A(w);\alpha] \geq [\eta_1, \eta_2;\alpha]$$

ii. $w \in [A;\alpha]_{[\eta_1,\eta_2;\alpha]}$,

$$[A(w);\alpha] \geq [\eta_1, \eta_2;\alpha] \Leftrightarrow A^L(w) \geq \eta_1 \text{and} A^U(w) \leq \eta_2$$

$$\Leftrightarrow w \in A^L_{\eta_1} \text{andw} \in A^U_{\eta_2}$$

$$\Leftrightarrow w \in A^L_{\eta_1} \cap A^U_{\eta_2}$$

iii. $w \in \left(\left[A \bigsqcup B;\alpha\right]\right)_{[\eta_1,\eta_2;\alpha]}$,

$$\left[\left(A \bigsqcup B\right)(w);\alpha\right] \geq [\eta_1, \eta_2;\alpha]$$

$$\Leftrightarrow \left[\sup\left\{A^L(w), B^L(w)\right\}, \inf\left\{A^U(w), B^U(w)\right\};\alpha\right] \geq [\eta_1, \eta_2;\alpha]$$

$$\Leftrightarrow \sup\left\{A^L(w), B^L(w)\right\} \geq \eta_1 \text{andinf}\left\{A^U(w), B^U(w)\right\} \leq \eta_2$$

$$\Leftrightarrow \left\{A^L(w) \geq \eta_1 \text{or} B^L(w) \geq \eta_1\right\} \text{and} \left\{A^U(w) \leq \eta_2 \text{or} B^U(w) \leq \eta_2\right\}$$

$$\Leftrightarrow \left\{A^L(w) \geq \eta_1 \text{and} A^U(w) \leq \eta_2\right\} \text{or} \left\{B^L(w) \geq \eta_1 \text{and} B^U(w) \leq \eta_2\right\}$$

$$\text{or}\left\{A^L(w) \geq \eta_1 \text{and} B^U(w) \leq \eta_2\right\} \text{or} \left\{B^L(w) \geq \eta_1 \text{and} A^U(w) \leq \eta_2\right\}$$

$$\Leftrightarrow w \in [A;\alpha]_{[\eta_1,\eta_2;\alpha]} \text{orw} \in [B;\alpha]_{[\eta_1,\eta_2;\alpha]} \text{or} \left\{w \in A^L_{\eta_1} \cap B^U_{\eta_2}\right\} \text{or} \left\{w \in B^L_{\eta_1} \cap A^U_{\eta_2}\right\}$$

$$\Leftrightarrow w \in [A;\alpha]_{[\eta_1,\eta_2;\alpha]} \cup [B;\alpha]_{[\eta_1,\eta_2;\alpha]} \cup \left(A^L_{\eta_1} \cap B^U_{\eta_2}\right) \cup \left(B^L_{\eta_1} \cap A^U_{\eta_2}\right)$$

iv. $w \in ([A \sqcap B;\alpha])_{[\eta_1,\eta_2;\alpha]}$,

$$[(A \sqcap B)(w); \alpha] \geq [\eta_1, \eta_2; \alpha]$$

$$\Leftrightarrow \left[\inf\left\{A^L(w), B^L(w)\right\}, \sup\left\{A^U(w), B^U(w)\right\}; \alpha\right] \geq [\eta_1, \eta_2; \alpha]$$

$$\Leftrightarrow \inf\left\{A^L(w), B^L(w)\right\} \geq \eta_1 \text{and} \sup\left\{A^U(w), B^U(w)\right\} \leq \eta_2$$

$$\Leftrightarrow \left\{A^L(w) \geq \eta_1 \text{and} B^L(w) \geq \eta_1\right\} \text{and} \left\{A^U(w) \leq \eta_2 \text{and} B^U(w) \leq \eta_2\right\}$$

$$\Leftrightarrow \left\{A^L(w) \geq \eta_1 \text{and} A^U(w) \leq \eta_2\right\} \text{and} \left\{B^L(w) \geq \eta_1 \text{and} B^U(w) \leq \eta_2\right\}$$

$$\Leftrightarrow w \in [A; \alpha]_{[\eta_1, \eta_2; \alpha]} \cap [B; \alpha]_{[\eta_1, \eta_2; \alpha]}$$

v. $w \in A^L_{\eta_2}$,

$$A^L(w) \geq \eta_2 \geq \eta_1 \Rightarrow w \in A^L_{\eta_1} \Rightarrow A^L_{\eta_1} \supseteq A^L_{\eta_2}$$

vi. $w \in A^U_{\eta_1}$ isgivenarbitrary.

$$A^U(w) \leq \eta_1 \leq \eta_2 \Rightarrow w \in A^U_{\eta_2} \Rightarrow A^U_{\eta_1} \subseteq A^U_{\eta_2}$$

vii. $w \in \bigcap_{i \in I} {A^L}_{\eta_i}$,

$$w \in \bigcap_{i \in I} {A^L}_{\eta_i} \Leftrightarrow \forall i \in I, w \in A^L_{\eta_i} \Leftrightarrow \forall i \in I, A^L(w) \geq \eta_i$$

$$\Leftrightarrow A^L(w) \geq \bigwedge_{i \in I} \eta_i \Leftrightarrow w \in {A^L}_{\bigwedge_{i \in I} \eta_i}$$

viii. $w \in \bigcup_{j \in I} {A^U}_{\eta_j}$,

$$w \in \bigcup_{j \in I} {A^U}_{\eta_j} \Leftrightarrow \exists j \in I, w \in A^U_{\eta_j} \Leftrightarrow \exists j \in I, A^U(w) \leq \eta_j$$

$$\Leftrightarrow A^U(w) \leq \bigvee_{j \in I} \eta_j \Leftrightarrow w \in {A^U}_{\bigvee_{j \in I} \eta_j}$$

Example 3: Let $X = \{k, l, m, n\}$.

$$[A; 0.5] = \left\{ \begin{array}{l} [< k, [0.2, 0.7] >; 0.5], [< l, [0.1, 0.8] >; 0.5], \\ [< m, [0.45, 0.6] >; 0.5], [< n, [0.35, 1] >; 0.5] \end{array} \right\}$$

[A; 0.5] is 0.5-IVFSs;

i. $[A; 0.5]_{[0;1]} = \{w \in XA^L(w) \geq 0 \text{and} A^U(w) \leq 1\} = \{k, l, m, n\} = X$
ii. $[A; 0.5]_{[0.15;0.6]} = \{w \in XA^L(w) \geq 0.15 \text{and} A^U(w) \leq 0.6\} = \{m\}$
iii. $[A; 0.5]_{[0,05;0.75]} = \{w \in XA^L(w) \geq 0.05 \text{and} A^U(w) \leq 0.75\} = \{k, m\}$
iv. $[A; 0.5]_{[0.5;1.0]} = \{w \in XA^L(w) \geq 0.5 \text{and} A^U(w) \leq 1.0\} = \varnothing$
v. $[A; 0.5]_{[0.1;0.8]} = \{w \in XA^L(w) \geq 0.1 \text{and} A^U(w) \leq 0.8\} = \{k, l, m\}$

Proposition 10: X is a universal set.

$[A; \alpha] \in \alpha-$ IVFS(X),

$$S \notin D(I_\alpha) \Rightarrow [A; \alpha]_S = \varnothing$$

Proof: Let $S \notin D(I_\alpha)$.

Suppose that $[A; \alpha]_S \neq \varnothing$,

$$\Rightarrow \exists w \in [A; \alpha]_S \Rightarrow [A(w); \alpha] \geq S$$

$$\Rightarrow A^L(w) \geq S^L, A^U(w) \leq S^U \Rightarrow \alpha \geq A^L(w) \geq S^L \text{and} \alpha \leq A^U(w) \leq S^U$$

$$\Rightarrow S \in D(I_\alpha)$$

is contradiction. Then,

$$[A; \alpha]_S = \varnothing$$

4 α-Interval Valued Fuzzy Subgroups

In consequences of discussions about previous chapter, algebraic structures on those sets are studied like that α-IVF subgroup.

$\alpha-$ IVFS(G,[A; α]) represents α-IVF subgroup on $\alpha-$ IVFS(G).

Definition 16: G is a group.

$[A; \alpha] \in \alpha-$ IVFS(G). w, z $\in$ G,

i. $[A(wz); \alpha] \geq \inf\{[A(w); \alpha], [A(z); \alpha]\}$
ii. $\left[A\left(w^{-1}\right); \alpha\right] \geq [A(w); \alpha]$

Example 4: $(\mathbb{Z}, +)$ abelian group,

$\alpha = \frac{1}{3}, A : \mathbb{Z} \to D\left(I_{\frac{1}{3}}\right),$

$$\left[A(0); \frac{1}{3}\right] = \left[\frac{1}{3}, \frac{1}{3}; \frac{1}{3}\right]$$

and

$$\left[A(k); \frac{1}{3}\right] = \begin{cases} \left[\frac{1}{4}, \frac{2}{3}; \frac{1}{3}\right]; kisevenandk \neq 0 \\ \left[\frac{1}{5}, \frac{5}{6}; \frac{1}{3}\right]; kisodd \end{cases}$$

$\frac{1}{3}$-interval valued fuzzy subgroup.

5 Solution

1. k, m $\in \mathbb{Z}$,
1. condition

k, m are even.

$$k = m \Rightarrow \left[\frac{1}{4}, \frac{2}{3}; \frac{1}{3}\right] = \left[\frac{1}{4}, \frac{2}{3}; \frac{1}{3}\right] \Rightarrow \left[A(k); \frac{1}{3}\right] = \left[A(m); \frac{1}{3}\right]$$

2. condition

k, m are odd.

$$k = m \Rightarrow \left[\frac{1}{5}, \frac{5}{6}; \frac{1}{3}\right] = \left[\frac{1}{5}, \frac{5}{6}; \frac{1}{3}\right] \Rightarrow \left[A(k); \frac{1}{3}\right] = \left[A(m); \frac{1}{3}\right]$$

$$\inf\left\{\left[\frac{1}{4}, \frac{2}{3}; \frac{1}{3}\right], \left[\frac{1}{5}, \frac{5}{6}; \frac{1}{3}\right]\right\} = \left[\inf\left\{\frac{1}{4}, \frac{1}{5}\right\}, \sup\left\{\frac{2}{3}, \frac{5}{6}\right\}; \frac{1}{3}\right] = \left[\frac{1}{5}, \frac{5}{6}; \frac{1}{3}\right]$$

1. condition

If k, m are even, then k + m is also even.

$$\left[A(k+m); \frac{1}{3}\right] = \left[\frac{1}{4}, \frac{2}{3}; \frac{1}{3}\right] \geq \inf\left\{\left[\frac{1}{4}, \frac{2}{3}; \frac{1}{3}\right], \left[\frac{1}{4}, \frac{2}{3}; \frac{1}{3}\right]\right\} = \left[\frac{1}{4}, \frac{2}{3}; \frac{1}{3}\right]$$
$$= \inf\left\{\left[A(k); \frac{1}{3}\right], \left[A(m); \frac{1}{3}\right]\right\}$$

2. condition

If k is even and m is odd, then k + m is odd.

$$\left[A(k+m); \frac{1}{3}\right] = \left[\frac{1}{5}, \frac{5}{6}; \frac{1}{3}\right] \geq \inf\left\{\left[\frac{1}{4}, \frac{2}{3}; \frac{1}{3}\right], \left[\frac{1}{5}, \frac{5}{6}; \frac{1}{3}\right]\right\} = \left[\frac{1}{5}, \frac{5}{6}; \frac{1}{3}\right]$$
$$= \inf\left\{\left[A(k); \frac{1}{3}\right], \left[A(m); \frac{1}{3}\right]\right\}$$

3. condition

If k is odd and m is even, then k + m is odd.
The proof is similar in <u>2. condition</u>.

4. condition

If k, m are odd, then k + m is even.

$$\left[A(k+m); \frac{1}{3}\right] = \left[\frac{1}{4}, \frac{2}{3}; \frac{1}{3}\right] \geq \inf\left\{\left[\frac{1}{5}, \frac{5}{6}; \frac{1}{3}\right], \left[\frac{1}{5}, \frac{5}{6}; \frac{1}{3}\right]\right\} = \left[\frac{1}{5}, \frac{5}{6}; \frac{1}{3}\right]$$
$$= \inf\left\{\left[A(k); \frac{1}{3}\right], \left[A(m); \frac{1}{3}\right]\right\}$$

1. condition

If k is even, then −k is even.

$$\left[A(-k); \frac{1}{3}\right] = \left[\frac{1}{4}, \frac{2}{3}; \frac{1}{3}\right] \geq \left[\frac{1}{4}, \frac{2}{3}; \frac{1}{3}\right] = \left[A(k); \frac{1}{3}\right]$$

2. condition

If k is odd, then −k is odd.

$$\left[A(-k); \frac{1}{3}\right] = \left[\frac{1}{5}, \frac{5}{6}; \frac{1}{3}\right] \geq \left[\frac{1}{5}, \frac{5}{6}; \frac{1}{3}\right] = \left[A(k); \frac{1}{3}\right]$$

Proposition 11: G is a group.

$\alpha - \text{IVFS}(G, [A; \alpha])$. $w \in G$,

i. $\left[A\left(w^{-1}\right);\alpha\right]=[A(w);\alpha]$
ii. $[A(e);\alpha]\geq[A(w);\alpha]$

Proof: $w\in G$,

i. $$\left[A\left(\left(w^{-1}\right)^{-1}\right);\alpha\right]\geq\left[A\left(w^{-1}\right);\alpha\right]\Rightarrow[A(w);\alpha]\geq\left[A\left(w^{-1}\right);\alpha\right]$$

$$\Rightarrow\left[A\left(w^{-1}\right);\alpha\right]=[A(w);\alpha]$$

ii. $[A(e);\alpha]=\left[A\left(ww^{-1}\right);\alpha\right]\geq\inf\left\{[A(w);\alpha],\left[A\left(w^{-1}\right);\alpha\right]\right\}=[A(w);\alpha]$

Proposition 12: G is a group.

$[A;\alpha]\in\alpha-$ IVFS(G). $w,z\in G$,

$$\alpha-\mathrm{IVFS}(G,[A;\alpha])\iff\left[A\left(wz^{-1}\right);\alpha\right]\geq\inf\{[A(w);\alpha],[A(z);\alpha]\}$$

Proof: $w,z\in G$,

$$"\Rightarrow"$$

$$\left[A\left(wz^{-1}\right);\alpha\right]\geq\inf\left\{[A(w);\alpha],\left[A\left(z^{-1}\right);\alpha\right]\right\}=\inf\{[A(w);\alpha],[A(z);\alpha]\}$$

$$"\Leftarrow"$$

i. $\left[A\left(w^{-1}\right);\alpha\right]=\left[A\left(ew^{-1}\right);\alpha\right]\geq\inf\{[A(e);\alpha],[A(w);\alpha]\}=[A(w);\alpha]$
ii. $[A(wz);\alpha]=\left[A\left(w\left(z^{-1}\right)^{-1}\right);\alpha\right]\geq\inf\left\{[A(w);\alpha],\left[A\left(z^{-1}\right);\alpha\right]\right\}$

$$\geq\inf\{[A(w);\alpha],[A(z);\alpha]\}$$

Proposition 13: G is a group.

$[A;\alpha],[B;\alpha]\in\alpha-$ IVFS(G),

$$\alpha-\mathrm{IVFS}(G,[A;\alpha])\text{and}\alpha-\mathrm{IVFS}(G,[B;\alpha])\Rightarrow\alpha-\mathrm{IVFS}(G,[A\sqcap B;\alpha])$$

Proof: $w,z\in G$,

$$[(A \sqcap B)(wz^{-1}); \alpha]$$

$$\geq [\inf\{\inf\{A^L(w), A^L(z)\}, \inf\{B^L(w), B^L(z)\}\},$$
$$\sup\{\sup\{A^U(w), A^U(z)\}, \sup\{B^U(w), B^U(z)\}\}; \alpha]$$

$$= [\inf\{\inf\{A^L(w), B^L(w)\}, \inf\{A^L(z), B^L(z)\}\},$$
$$\sup\{\sup\{A^U(w), B^U(w)\}, \sup\{A^U(z), B^U(z)\}\}; \alpha]$$

$$= \inf\{[\inf\{A^L(w), B^L(w)\}, \sup\{A^U(w), B^U(w)\}; \alpha],$$
$$[\inf\{A^L(z), B^L(z)\}, \sup\{A^U(z), B^U(z)\}; \alpha]\}$$

$$= \inf\{[(A \sqcap B)(w); \alpha], [(A \sqcap B)(z); \alpha]\}$$

Proposition 14: G is a group.

I is an index set $\ni i \in I$. $[A_{i\in I}; \alpha] \in \alpha-$ IVFS(G),

$$\alpha - \text{IVFS}(G, [A_{i\in I}; \alpha]) \Rightarrow \alpha - \text{IVFS}(G, [\sqcap_{i\in I} A_i; \alpha])$$

Proof: $w, z \in G$,

$$[(\cap_{i\in I} A_i)(wz^{-1});\ \alpha] = \left[\bigwedge_{i\in I} A_i^L(wz^{-1}),\ \bigvee_{\in I} A_i^U(wz^{-1}); \alpha\right]$$

$$\geq \left[\bigwedge_{i\in I} \inf\{A^L{}_i(w), A^L{}_i(z)\}, \bigvee_{i\in I} \sup\{A^U{}_i(w), A^U{}_i(z)\}; \alpha\right]$$

$$= \left[\inf\left\{\bigwedge_{i\in I} A^L{}_i(w), \bigwedge_{i\in I} A^L{}_i(z)\right\}, \sup\left\{\bigvee_{i\in I} A^U{}_i(w), \bigvee_{i\in I} A^U{}_i(z)\right\}; \alpha\right]$$

$$= \inf\left\{\left[\bigwedge_{i\in I} A^L{}_i(w), \bigvee_{i\in I} A^U{}_i(w); \alpha\right], \left[\bigwedge_{i\in I} A^L{}_i(z), \bigvee_{i\in I} A^U{}_i(z); \alpha\right]\right\}$$

$$\Rightarrow \alpha - \text{IVFS}(G, [\sqcap_{i\in I} A_i; \alpha])$$

Proposition 15: G is a group.

$$\alpha - \mathrm{IVFS}(G, [A; \alpha]) \iff \forall[\eta_1, \eta_2; \alpha] \in D(I_\alpha)$$

$$[A; \alpha]_{[\eta_1,\eta_2;\alpha]} \neq \varnothing, [\eta_1, \eta_2; \alpha] - \text{levelsubsetof}[A; \alpha]\text{issubgroupofG}.$$

Proof: " $\Rightarrow$ "

i. $w \in G, \exists[\eta_1, \eta_2; \alpha] \in D(I_\alpha), [A(w); \alpha] = [\eta_1, \eta_2; \alpha] \Rightarrow \forall w \in G, [A(e); \alpha] \geq [A(w); \alpha]$

$$\Rightarrow e \in [A; \alpha]_{[\eta_1,\eta_2;\alpha]} \Rightarrow [A; \alpha]_{[\eta_1,\eta_2;\alpha]} \neq \varnothing$$

$w, z \in [A; \alpha]_{[\eta_1,\eta_2;\alpha]}$

$$[A(w); \alpha] \geq [\eta_1, \eta_2; \alpha]\text{and}[A(z); \alpha] \geq [\eta_1, \eta_2; \alpha]$$

$$\Rightarrow \left[A\left(wz^{-1}\right); \alpha\right] \geq \inf\{[A(w); \alpha], [A(z); \alpha]\} \geq [\eta_1, \eta_2; \alpha] \Rightarrow wz^{-1} \in [A; \alpha]_{[\eta_1,\eta_2;\alpha]}$$

" $\Leftarrow$ "

Suppose $\exists w_0, z_0 \in G, \left[A\left(w_0 z_0{}^{-1}\right); \alpha\right] < [\inf\{A(w_0), A(z_0); \alpha\}]$

$$[A(w_0); \alpha] = [a_1, a_2; \alpha], [A(z_0); \alpha] = [b_1, b_2; \alpha]\text{istaken},$$

$$[\eta_1, \eta_2; \alpha] = \inf\{[a_1, a_2; \alpha], [b_1, b_2; \alpha]\}\text{ischosen},$$

$$[A(w_0); \alpha] = [a_1, a_2; \alpha] \geq \inf\{[a_1, a_2; \alpha], [b_1, b_2; \alpha]\} = [\eta_1, \eta_2; \alpha] \Rightarrow w_0 \in [A; \alpha]_{[\eta_1,\eta_2;\alpha]}$$
$$[A(z_0); \alpha] = [b_1, b_2; \alpha] \geq \inf\{[a_1, a_2; \alpha], [b_1, b_2; \alpha]\} = [\eta_1, \eta_2; \alpha] \Rightarrow z_0 \in [A; \alpha]_{[\eta_1,\eta_2;\alpha]}$$

$$\left[A\left(w_0 z_0{}^{-1}\right); \alpha\right] < \inf\{[A(w_0); \alpha], [A(z_0); \alpha]\} = \inf\{[a_1, a_2; \alpha], [b_1, b_2; \alpha]\}$$

$$= [\eta_1, \eta_2; \alpha] \Rightarrow w_0 z_0{}^{-1} \notin [A; \alpha]_{[\eta_1,\eta_2;\alpha]}$$

is contradiction. Then,

$$\forall w_0, z_0 \in G, \left[A\left(w_0 z_0{}^{-1}\right); \alpha\right] \geq \inf\{[A(w_0); \alpha], [A(z_0); \alpha]\}$$

Proposition 16: G is a group.

$\alpha - \mathrm{IVFS}(G, [A; \alpha])$ and $[\eta_1, \eta_2; \alpha] \in D(I_\alpha)$,

$$[A; \alpha]_{[\eta_1,\eta_2;\alpha]}\text{issubgroupofG} \iff A^L_{\eta_1}\text{ve}\left(\left(A^U\right)^c\right)_{1-\eta_2}\text{aresubgroupsofG}.$$

Proof: " ⇒ "

$$\forall[\eta_1, \eta_2; \alpha] \in D(I_\alpha), e \in [A; \alpha]_{[\eta_1,\eta_2;\alpha]} = A^L_{\eta_1} \cap A^U_{\eta_2} \Rightarrow e \in A^L_{\eta_1} \mathrm{vee} \in A^U_{\eta_2}$$

$$\Rightarrow A^U(e) \leq \eta_2 \Rightarrow 1 - A^U(e) \geq 1 - \eta_2 \Rightarrow e \in \left(\left(A^U\right)^c\right)_{1-\eta_2}$$

$$A^L_{\eta_1} \neq \varnothing \mathrm{and}\left(\left(A^U\right)^c\right)_{1-\eta_2} \neq \varnothing$$

$w, z \in A^L_{\eta_1}$,

$$A^L(w) \geq \eta_1 \mathrm{and} A^L(z) \geq \eta_1$$

$$\Rightarrow A^L\left(wz^{-1}\right) \geq \inf\left\{A^L(w), A^L(z)\right\} \geq \eta_1 \Rightarrow wz^{-1} \in A^L_{\eta_1}$$

$w, z \in \left(\left(A^U\right)^c\right)_{1-\eta_2}$,

$$1 - A^U(w) \geq 1 - \eta_2 \mathrm{and} 1 - A^U(z) \geq 1 - \eta_2 \mathrm{and}$$

$$A^U\left(wz^{-1}\right) \leq \sup\left\{A^U(w), A^U(z)\right\}$$

$$\Rightarrow 1 - A^U\left(wz^{-1}\right) \geq 1 - \sup\left\{A^U(w), A^U(z)\right\}$$

$$\Rightarrow 1 - A^U\left(wz^{-1}\right) \geq \inf\left\{1 - A^U(w), 1 - A^U(z)\right\} \geq 1 - \eta_2$$

$$\Rightarrow wz^{-1} \in \left(\left(A^U\right)^c\right)_{1-\eta_2}$$

" ⇐ "

$$A^L_{\eta_1} \mathrm{and}\left(\left(A^U\right)^c\right)_{1-\eta_2} \mathrm{aresubgroupsofG} \Rightarrow e \in A^L_{\eta_1} \mathrm{and} e \in \left(\left(A^U\right)^c\right)_{1-\eta_2}$$

$$\Rightarrow 1 - A^U(e) \geq 1 - \eta_2 \Rightarrow A^U(e) \leq \eta_2 \Rightarrow e \in A^U_{\eta_2}$$

$$\Rightarrow e \in A^L_{\eta_1} \cap A^U_{\eta_2} = [A; \alpha]_{[\eta_1,\eta_2;\alpha]} \Rightarrow [A; \alpha]_{[\eta_1,\eta_2;\alpha]} \neq \varnothing$$

$w, z \in G,$

$$\Rightarrow A^L\left(wz^{-1}\right) \geq \inf\left\{A^L(w), A^L(z)\right\} \geq \eta_1 \text{and}$$

$$1 - A^U\left(wz^{-1}\right) \geq \inf\left\{1 - A^U(w), 1 - A^U(z)\right\} \geq 1 - \eta_2$$

$$\Rightarrow A^L\left(wz^{-1}\right) \geq \inf\left\{A^L(w), A^L(z)\right\} \geq \eta_1 \text{and}$$

$$1 - A^U\left(wz^{-1}\right) \geq 1 - \sup\left\{A^U(w), A^U(z)\right\} \geq 1 - \eta_2$$

$$\Rightarrow A^L\left(wz^{-1}\right) \geq \inf\left\{A^L(w), A^L(z)\right\} \geq \eta_1 \text{and} A^U\left(wz^{-1}\right) \leq \sup\left\{A^U(w), A^U(z)\right\} \leq \eta_2$$

$$\Rightarrow \left[A^L\left(wz^{-1}\right), A^U\left(wz^{-1}\right); \alpha\right] \geq [\eta_1, \eta_2; \alpha]$$

$$\Rightarrow wz^{-1} \in [A; \alpha]_{[\eta_1, \eta_2; \alpha]}$$

Example 5: $(\mathbb{Z}, +)$ abelian group,

$\alpha = \frac{1}{3}, A : \mathbb{Z} \to D\left(I_{\frac{1}{3}}\right),$

$$\left[A(0); \frac{1}{3}\right] = \left[\frac{1}{3}, \frac{1}{3}; \frac{1}{3}\right]$$

and

$$\left[A(k); \frac{1}{3}\right] = \begin{cases} \left[\frac{1}{4}, \frac{2}{3}; \frac{1}{3}\right]; kisevenandk \neq 0 \\ \left[\frac{1}{5}, \frac{5}{6}; \frac{1}{3}\right]; kisodd \end{cases}$$

$\frac{1}{3}$-IVF subgroup,

$$\left[\frac{1}{6}, \frac{4}{5}; \frac{1}{3}\right] \in D\left(I_{\frac{1}{3}}\right)$$

$$\left[A; \frac{1}{3}\right]_{\left[\frac{1}{6}, \frac{4}{5}; \frac{1}{3}\right]} = 2\mathbb{Z} \text{issubgroupof} \mathbb{Z}$$

$$A^L{}_{\frac{1}{6}} = \mathbb{Z}\text{and}\left(\left(A^U\right)^c\right)_{1-\frac{4}{5}} = \left(\left(A^U\right)^c\right)_{\frac{1}{5}} = 2\mathbb{Z}\text{aresubgroupsof}\mathbb{Z}$$

Definition 17: X is a universal set.

$[A; \alpha] \in \alpha-$ IVFS(X),

$$A^* = \{w \in X[A(w); \alpha] > [0, 1; \alpha]\}$$

is named support set of A.

Definition 18: G is a group.
$\alpha - \mathrm{IVFS}(G, [A; \alpha])$

$$A_* = \{w \in G[A(w); \alpha] = [A(e); \alpha]\}$$

Proposition 17: G is a group.

$\alpha - \mathrm{IVFS}(G, [A; \alpha])$,

i. A^* is subgroup of G.
ii. A_* is subgroup of G.

Proof:

i. $w, z \in A^*$

$$[A(w); \alpha]\mathrm{and}[A(z); \alpha] > [0, 1; \alpha] \Rightarrow \inf\{[A(w); \alpha], [A(z); \alpha]\} > [0, 1; \alpha]$$

$$\Rightarrow \left[A\left(wz^{-1}\right); \alpha\right] \geq \inf\{[A(w); \alpha], [A(z); \alpha]\} > [0, 1; \alpha] \Rightarrow wz^{-1} \in A^*$$

ii. $w, z \in A_*$,

$$\inf\{[A(w); \alpha], [A(z); \alpha]\} = [A(e); \alpha]$$

$$\Rightarrow \left[A\left(wz^{-1}\right); \alpha\right] \geq \inf\{[A(w); \alpha], [A(z); \alpha]\} = [A(e); \alpha] \ldots (1)$$

ontheotherhand

$$\alpha - \mathrm{IVFS}(G, [A; \alpha]) \Rightarrow [A(e); \alpha] \geq \left[A\left(wz^{-1}\right); \alpha\right] \ldots (2)$$

from inequalities above,

$$\left[A\left(wz^{-1}\right); \alpha\right] = [A(e); \alpha] \Rightarrow wz^{-1} \in A_*$$

6 Conclusion

In the paper, the elemental properties of α-IVSs are studied. It is proved that α-IVSs is lattice by using order relation on these sets.

After the studies on α-IVSs, the concept of α-IVFSs are introduced. The essential algebraic statements such as intersection and union on these sets are given. Another important point for set theory is complement of a set. For this reason, the complement of α-IVFSs is given by using of negation function on α-IVFSs. It is studied that Morgan Laws are satisfied on these sets. In order to determine the relation between α-IVFSs and crisp sets, the concept and basic points of level subset of α-IVFSs are given.

After the studies on α-IVFSs, the definition of α-IVF subgroups is introduced. Structural properties of these subgroups are studied. The relations between subgroups of classical G group and the level subsets of α-IVFSs of classical G groups are studied.

References

1. L.A. Zadeh, Fuzzy Sets. Inform. Control **8**(3), 338–353 (1965)
2. L.A. Zadeh, The concept of a linguistic variable and its application to approximate reasoning. Part 1, Infor. Sci. **8**(3), 199–249 (1975)
3. L.A. Zadeh, The concept of a linguistic variable and its application to approximate reasoning. Part 2, Infor. Sci. **8**(4), 301–357 (1975)
4. L.A. Zadeh, The concept of a linguistic variable and its application to approximate reasoning. Part 3, Infor. Sci. **9**(1), 43–80 (1975)
5. I. Grattan-Guiness, Fuzzy membership mapped onto interval and many-valued quantities. Z.Math. Logik, Grundladen Math. **22**(1), 149–160 (1975)
6. B. Gorzalczany, Approximate inference with interval-valued fuzzy sets, in: Proceeding Polish Symp. on Interval and Fuzzy Mathematics, Poznan, pp. 89–95 (1983)
7. B. Gorzalczany, A method of inference in approximate reasoning based on interval-valued fuzzy sets. Fuzzy Sets Syst. **21**(1), 1–17 (1987)
8. K.U. Jahn, Intervall-wertige Mengen. Math. Nach. **68**(1), 115–132 (1975)
9. R. Sambuc, Fonctions φ-floues application L'aide au diagnostic en pathologie Thyroidi- enne, Ph.D. Thesis, Univ. Marseille, France (1975)
10. I. Turksen, Interval valued fuzzy sets based on normal forms. Fuzzy Sets Syst **20**(2), 191–210 (1986)
11. T.K. Mondal, S.K. Samantha, Topology of interval-valued fuzzy sets. Indian J. Pure Applied Math. **30**(1), 20–38 (1999)
12. A. Rosenfeld, Fuzzy groups. J. Mathem. Analy. Applic. **35**(3), 512–517 (1971)
13. R. Biswas, Rosenfeld's fuzzy subgroups with interval valued membership functions. Fuzzy Sets Syst. **63**(1), 87–90 (1994)
14. H.W. Kang, K. Hur, Interval-valued fuzzy subgroups and rings. Honam Mathem. J. **32**(4), 593–617 (2010)
15. G. Wang, X. Li, TH-Interval Valued Fuzzy Subgroups. J. Lanzhou University **32**, 96–99 (1996)
16. X. Li, G. Wang, The SH-interval-valued fuzzy subgroups. Fuzzy Sets Syst. **112**(2), 319–325 (2000)

Design of a Fuzzy PID Controller for the Balancing Problem of a Quadruped Robot

Jose Eduardo Lopez-Ramos, Oscar Castillo, and Patricia Melin

1 Introduction

Quadruped robots are machines designed to imitate the movement of four leg animals. These robots have four extremities, each of them with their own sensors and actuators, which enable a movement similar to quadruped animals, such as dogs, cats or horses [1–4]. Quadruped robots have been receiving increasing attention in the robotics area because they can have many real applications, such as search and rescue, exploration, agriculture and entertainment [5–7].

Balancing control is a key issue for quadruped robots, as it allows them to maintain stability and efficiently navigate in different terrains. Quadruped robots are designed with the goal of imitating real movement of quadruped animals, and require appropriate control algorithms for achieving stability and equilibrium while moving [8–10]. There are several kinds of quadruped robots in the literature, but we selected one that has not been previously considered with fuzzy logic. This robot is called Robot Dog Kit FNK0050 from Freenove, for which only a proportional-integral-derivative (PID) control existed in the previous literature. In this sense, this is an existing research gap in the state of the art that we decided to consider as our research work.

The contribution of this paper is the proposed design of fuzzy PID for the balancing problem as an enhancement to the previous existing PID controller. The main idea is that fuzzy logic enables having a nonlinear control model, which can provide better results for complex problems. In this way, fuzzy PID control will surpass the traditional linear PID control.

J. E. Lopez-Ramos · O. Castillo (✉) · P. Melin
Tijuana Institute of Technology, TecNM, Tijuana, Mexico
e-mail: ocastillo@tectijuana.mx

P. Melin
e-mail: pmelin@tectijuana.mx

P. Melin and O. Castillo (eds.), *New Directions on Hybrid Intelligent Systems Based on Neural Networks, Fuzzy Logic, and Optimization Algorithms*, Studies in Computational Intelligence 1146, https://doi.org/10.1007/978-3-031-53713-4_5

The rest of the document is structured as: Sect. 2 explains the quadruped robot utilized in this work. Section 3 outlines the existing PID controller for this robot and then the proposal for a fuzzy PID controller. Section 4 offers the conclusions and future work.

2 Quadruped Robots and Problem Definition

In this section we are describing the particular quadruped robot utilized in this research work. The Robot Dog Kit FNK0050 is one of the models from Freenove, which is quadruped design with open code that is compatible with Raspberry Pi. It is formed with an acrylic light structure with several sensors, such as camera, ultrasound sensor, gyroscope and accelerometer. The robot has in total 12 degrees of freedom, which makes it an ideal system for low cost robotic applications. Figure 1 shows a side view the robot dog after it was assembled. Figure 2 depicts a front view of the robot.

Controlling the stability of quadruped robots is very important for several reasons: stability will allow the robot without falling, robustness against external perturbations, energy efficiency and improving performance.

Fig. 1 Robot Dog Kit side view

Fig. 2 Robot Dog Kit front view

3 Proposed Fuzzy PID

In this section the structure of the existing PID controller for the robot is presented, as well as the proposed fuzzy PID for improving the robot balancing and stability.

First, we briefly describe the existing PID control method for balancing the robot dog. This controller is based on three components: proportional (P), integral (I) and derivative (D). The sum of the three controllers will produce the total control applied to the robot. This is illustrated in Fig. 3.

Now we are presenting the fuzzy PID controller for the balancing problem of the quadruped robot. In this case, we now have a fuzzy logic controller (FLC), which is a new block in the diagram of Fig. 4. The idea of having a FLC is for determining the appropriate gain values that will be utilized by the PID controller. In other words, we will have a set of fuzzy rules that will dynamically change the gains used in the PID block, in this way achieving flexibility to different conditions.

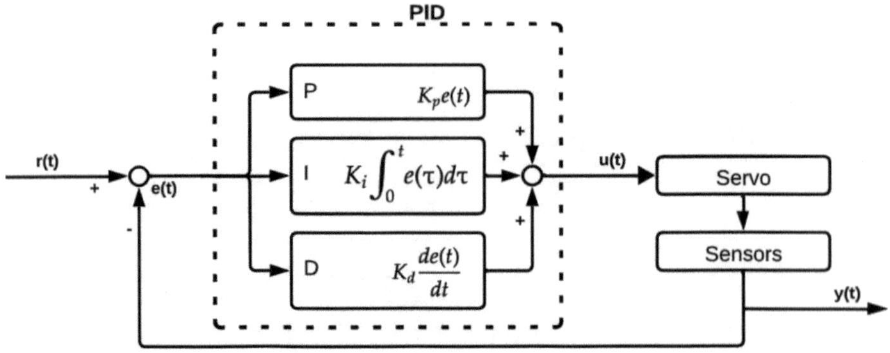

Fig. 3 Existing PID controller for balancing the robot

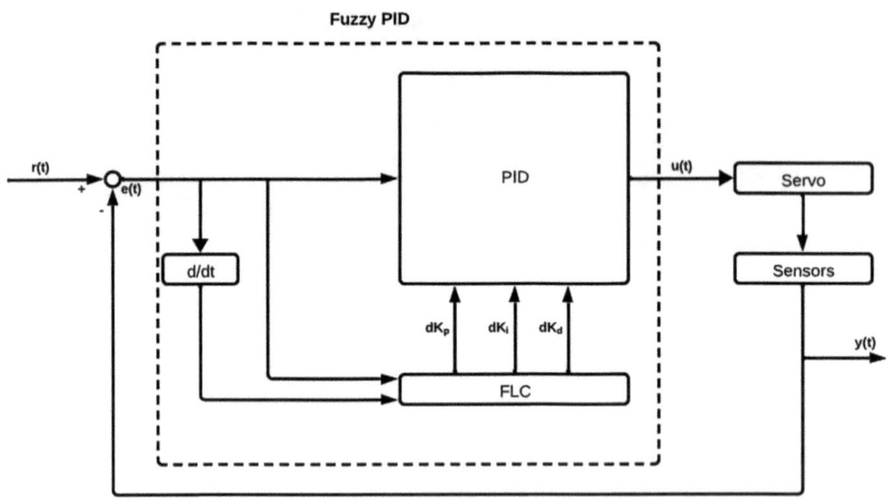

Fig. 4 Proposed Fuzzy PID controller for balancing the robot

There are previous works, where the fuzzy PID approach has already been used in quadruped robots with good results [8–11], but this has not been done with this particular robot. We expect to also achieve good results with the Robot Dog Kit.

4 Conclusions

In this paper we presented fuzzy PID control for the balancing problem of a quadruped robot. The existing PID controller was enhanced with fuzzy logic to improve its performance. The balancing problem is very important to achieve efficient movement of the robot, as well as to minimize energy usage. A comparison of the designed fuzzy

PID controller with respect to the existing PID controller was presented to verify the superiority of the proposal. The experimentation was performed with a particular robot called FNK0050 for which only a PID controller was previously used, so the contribution is in the enhancement of the control using fuzzy logic with the goal of achieving better results. Future works include optimizing the design of the fuzzy PID controller, as well as elevating its design to type-2 [12, 13] and type-3 fuzzy logic [14–16] with the goal of handling uncertainty, in the control process, in a better way. Finally, we envision optimizing the fuzzy PID design with the utilization of metaheuristics, such as in [17–21].

References

1. J.T. Machado, M.F. Silva, An overview of legged robots. In International symposium on mathematical methods in engineering (Vol. 48, No. 2.3). Ankara, Turkey: MME Press (2006)
2. P.G. De Santos, E. Garcia, J. Estremera, *Quadrupedal locomotion: an introduction to the control of four-legged robots*, vol. 1 (Springer, London, 2006)
3. S. Hirose, Y. Fukuda, K. Yoneda, A. Nagakubo, H. Tsukagoshi, K. Arikawa, ... R. Hodoshima, Quadruped walking robots at Tokyo Institute of Technology. IEEE Robot. Autom. Mag. **16**(2), 104-114 (2009)
4. J. Ingvast, C. Ridderström, J. Wikander, The four legged robot system WARP1 and its capabilities. In Second Swedish Workshop on Autonomous Systems (2002)
5. M.P. Murphy, A. Saunders, C. Moreira, A.A. Rizzi, M. Raibert, The littledog robot. Intern. J. Rob. Res. **30**(2), 145–149 (2011)
6. B. Katz, J. Di Carlo, S. Kim, Mini cheetah: A platform for pushing the limits of dynamic quadruped control. In 2019 international conference on robotics and automation (ICRA) (pp. 6295–6301). IEEE (2019)
7. P. Biswal, P.K. Mohanty, Development of quadruped walking robots: a review. Ain Shams Eng. J. **12**(2), 2017–2031 (2021)
8. K. Arphakorn, P. Amornphun, J. Wisanu, Gait control of a four-legged robot with fuzzy-PID controller. Proceedings of International Conference on Artificial Life and Robotics (Vol. 25, pp. 514–518) (2020)
9. K. Chang, X.J. Han, Y. Yang, Self-adaptive pid control of hydraulic quadruped robot. Appl. Mech. Mater. **496**, 1407–1412 (2014)
10. K.Y. Chen, C.Y. Tsui, The fuzzy control approach for a quadruped robot guide dog. Int. J. Fuzzy Syst. **23**, 1789–1796 (2021)
11. O. Montiel, R. Sepulveda, P. Melin, O. Castillo, M.A. Porta García, I.M. Meza-Sánchez, Performance of a simple Tuned Fuzzy controller and a PID controller on a DC Motor. FOCI 2007: 531–537 (2020)
12. T. Kumbasar, H. Hagras, Interval type-2 Fuzzy PID controllers. In: Kacprzyk, J., Pedrycz, W. (eds) Springer Handbook of Computational Intelligence. Springer Handbooks. Springer, Berlin, Heidelberg (2015)
13. P. Melin, O. Castillo, A new method for adaptive control of non-linear plants using type-2 fuzzy logic and neural networks. Int. J. Gen. Syst. **33**(2–3), 289–304 (2004)
14. A. Mohammadzadeh, M.H. Sabzalian, W. Zhang, An interval type-3 fuzzy system and a new online fractional-order learning algorithm: theory and practice. IEEE Trans. Fuzzy Syst. **28**(9), 1940–1950 (2020)
15. O. Castillo, J.R. Castro, P. Melin, *Interval Type-3 fuzzy systems: theory and design* (Springer, Cham, Switzerland, 2022)
16. O. Castillo, J.R. Castro, P. Melin, Interval type-3 fuzzy control for automated tuning of image quality in televisions. Axioms **11**, 276 (2022)

17. F. Valdez, P. Melin, O. Castillo, Evolutionary method combining particle swarm optimization and genetic algorithms using fuzzy logic for decision making, in: IEEE International Conference on Fuzzy Systems, pp. 2114–2119 (2009)
18. F. Valdez, J.C. Vazquez, P. Melin, O. Castillo, Comparative study of the use of fuzzy logic in improving particle swarm optimization variants for mathematical functions using co-evolution. Appl. Soft Comput. **52**, 1070–1083 (2017)
19. D. Sanchez, P. Melin, O. Castillo, A grey wolf optimizer for modular granular neural networks for human recognition. Comput. Intell. Neurosci. **2017**(2017). https://doi.org/10.1155/2017/4180510
20. O. Castillo, E. Lizzarraga, J. Soria, P. Melin, F. Valdez, New approach using ant colony optimization with ant set partition for fuzzy control design applied to the ball and beam system. Inf. Sci. **294**, 203–215 (2015)
21. L. Amador-Angulo, O. Mendoza, J.R. Castro, A. Rodriguez-Diaz, P. Melin, O. Castillo, Fuzzy sets in dynamic adaptation of parameters of a bee colony optimization for controlling the trajectory of an autonomous mobile robot. Sensors **16**(9), 1458 (2016)

Neural Networks

Modular Perspective for Population and Gross National Income Time Series Prediction Using a Neural Network Model: A Case Study of OECD Member Countries

Martha Ramirez and Patricia Melin

1 Introduction

A time series consists of a set of observations that are chronologically ordered, and frequently they are used to represent the change of a variable over time during a given period. This variable can be statistical, demographic, financial, or industrial, among others. In general, a set of available observations is found stored at equal time intervals.

The objective of the time series analysis is the knowledge of its behavior pattern, to be able to notice its evolution in the near future, assuming that the conditions will not vary significantly. Although the behavior of any time series can be graphically observed, not in all cases it is possible to distinguish the particularities that each one can present.

There are certain characteristic movements or variations that can be separately measured and observed. These movements are often called components of a time series and are assumed to be caused by different phenomena. In other words, they contemplate a significant part within the analysis of historical information.

During the last decades, many statistical models have been used to predict time series [1, 2], being Artificial Neural Networks (ANNs) one of the most robust and efficient methods due to its non-linear prediction advantages [3, 4], which with their technological advances have contributed to solve complex problems, and this is due to the ability of ANNs to learn non-linear relationships [5–7].

Supervised ANNs are frequently used due to their effectiveness in performing time series prediction tasks, since they can produce desired results in response to

M. Ramirez · P. Melin (✉)
Tijuana Institute of Technology, TecNM, Tijuana, Mexico
e-mail: pmelin@tectijuana.mx

P. Melin and O. Castillo (eds.), *New Directions on Hybrid Intelligent Systems Based on Neural Networks, Fuzzy Logic, and Optimization Algorithms*, Studies in Computational Intelligence 1146, https://doi.org/10.1007/978-3-031-53713-4_6

sample inputs [8–10]. Unlike unsupervised ANNs that only need to have input data, making it possible clustering time series even when the outputs associated with them are unknown [11–13].

Therefore, a novel ANN method for predicting the time series of population and gross national income (GNI), in which both supervised and unsupervised neural networks are incorporated, is presented.

Thus, with our proposal the main contribution is a method that contemplates, in the first instance, time series prediction using supervised neural networks (monolithic and modular) and an unsupervised neural network responsible for generating data clusters that represent the input of a supervised modular network. Subsequently, the results of the neural networks are integrated to improve the quality of the obtained prediction.

It is important to mention that the main difference of our proposal for time series prediction with multiple variables is that independent results are obtained for each neural network and subsequently seeks to improve the quality of the prediction by integrating the results of all the networks. In addition to using both supervised and unsupervised networks in the model.

Beneath we briefly describe the contents of this paper. Section 2 offers an overview of the problem. We describe the used methods in Sect. 3. The experimental work and discussion results are presented in Sects. 4 and 5, respectively. Lastly, in Sect. 6, the conclusions are presented.

2 Problem Description

The Organization for Economic Co-operation and Development (OECD) is a unique forum, a knowledge center for data collection and analysis, exchange of experience and good practice, where the policies and the dissemination of best practices in areas such as investment, education, inclusion, competition, good governance, among others, have been promoted to design better policies for a better life.

Its efforts are focused on identifying the factors that lead to positive economic, social, and environmental changes, for which it sets international standards. Currently the organization has 38 member countries (Table 1).

The growth or decrease in the size of the human population can significantly impact a country in the use of natural resources, social infrastructure and availability of goods and services, among others.

The development of a country contemplates social, environmental, and economic aspects or indicators, among others. Therefore, multiple measurements are required to obtain a complete picture. In the specific case of the total income of the residents of an economy, measured by their gross national income (GNI), it could be said that it is a good measure of the capacity of a country to ensure the well-being of its people.

Meanwhile, starting from the premise that it is difficult to select a reliable prediction method, since the vast majority can sometimes be more or less accurate depending on the specific circumstances and the complexity of the data sets

Table 1 OECD member countries

Name	Code	Name	Code	Name	Code	Name	Code
Luxembourg	LUX	Colombia	COL	Korea, Rep	KOR	Belgium	BEL
Mexico	MEX	Costa Rica	CRI	Latvia	LVA	Canada	CAN
Netherlands	NLD	Czech Republic	CZE	Lithuania	LTU	Denmark	DNK
New Zealand	NZL	Germany	DEU	Norway	NOR	Estonia	EST
Slovenia	SVN	Greece	GRC	Poland	POL	Finland	FIN
Spain	ESP	Hungary	HUN	Portugal	PRT	France	FRA
Sweden	SWE	Iceland	ISL	Slovak Republic	SVK	Ireland	IRL
Switzerland	CHE	Italy	ITA	Turkey	TUR	Israel	ISR
United States	USA	Australia	AUS	United Kingdom	GBR		
Chile	CHL	Japan	JPN	Austria	AUT		

available, it is necessary to have tools that make it possible to predict the population and gross national income (GNI) growth, in order to analyze the impact of their change over time [14].

For this work for each of the OECD member countries we use a data set consisting of two time series: population (1960–2020) [15] and gross national income (2006–2020) [16]. It should be noted that both series have six attributes each (country name, country code, indicator name, indicator code, value indicator, year) and no data preprocessing was performed.

3 Methodology

In this paper, we propose a model consisting of using multiple neural networks to improve the prediction quality for a given set of time series.

For clustering tasks, we use competitive networks to find similar data, form clusters where each element is labeled with the class to which it belongs. This means, all input units $\boldsymbol{i}$ are connected to all output units $\boldsymbol{j}$ with weight $\boldsymbol{w_{ij}}$. The input dimension is the number of inputs, and the total number of outputs is the same as the number of clusters into which the data will be segmented.

The activation value $\boldsymbol{a_j}$ of the output unit is calculated by the inner product of the input and weight vectors and returns the output unit with the highest activation $\boldsymbol{k}$, the weights leading to this unit are updated by the competitive function or the so-called winner-take-all learning rule as Eq. (1) and the weights of the output unit with the smallest activation are updated to be even closer, and can be written as Eq. (2) [17]:

$$w_k(t+1) = \frac{w_k(t) + \eta(x(t) - w_k(t))}{\|w_k(t) + \eta(x(t) - w_k(t))\|} \quad (1)$$

$$w_k(t+1) = w_k(t) + \eta(x(t) - w_k(t)) \quad (2)$$

where w is an input weight vector, k is the winner output unit, t is a vector position, η is the learning rate and x is an input vector. Thus, when an input x is presented to the network, the weight vector closest to x rotates toward it.

The main objective is to detect the similarities in the dataset [18]. So, we have used the competitive neural network to separate the dataset into six classes: class 1(CS1), class 2 (SC2), class 3 (CS3), class 4 (CS4), class 5 (CS5) and the class 6 (CS6).

For the prediction tasks we used a Nonlinear Autoregressive (NAR) neural network [19, 20], because of the good results obtained previously.

By using a NAR Network, the future values of a time series y(t) are predicted only from past values of that series and can be written as Eq. (3) [21]:

$$y(t) = f(y(t-1), y(t-2), \ldots, y(t-d)) \quad (3)$$

where *y(t)* represents a value of the time series *y* at time *t*, *d* is a time delay parameter and *f* is an activation function.

We offer a computational model (Fig. 1) for time series prediction, which contemplates that each time series enters to three modules of a neural network. The first of which consists of a supervised monolithic NAR network, the second of a modular NAR network and the third of a competitive network that segments the information data to obtain clusters of information, then each data cluster enters a module of a modular NAR network.

Once the results of all the modules are obtained, they are integrated to improve the quality of the prediction. If the result after the integration does not improve the best prediction made by all the modules of the neural network, the best result is taken (the winning module).

It is important to point out that by having independent predictions from each module, it is possible to compare them with the result obtained by the integration, which consists of calculating the average of the predictions made by all the modules of the neural network. With the above, it is possible to take the best result accordingly.

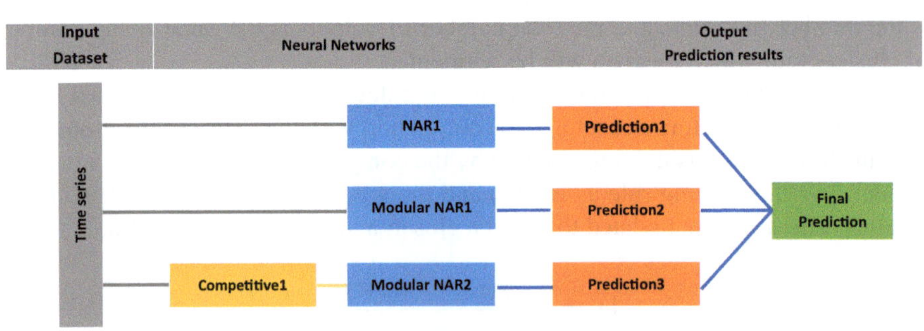

Fig. 1 Visual description of the proposed method

For prediction tasks a NAR network is used, where the hidden layer consists of 10 neurons and for training the Levenberg–Marquardt algorithm is used. For the modular NAR we used the same neural network architecture for each module. To perform the clustering tasks a competitive neural network of six classes are used (CS1, CS2, CS3, CS4, CS5 and CS6).

In this case we limit the number of clusters that we seek to obtain to six, however, it may be that the neural network forms a smaller number of clusters). Once the clusters are obtained, for each data segment we perform the prediction using a modular NAR neural network with the same architecture described above, where each module is responsible of performing the prediction for every cluster.

4 Experiments and Results

During the experimentation we use the entire data set, and in the case of supervised neural networks 70% was used for training and 30% for validation and testing. Each experiment consists of performing 30 executions for both supervised and unsupervised neural networks.

For clustering, we use competitive neural networks. In the case of the GNI time series, three classes were obtained: the CS1 class with 32 elements, the CS2 class with 5 elements and finally the CS3 class with one element (Table 2).

In the case of the population time series, four classes were obtained: the CS1 class with 26 elements, the CS2 class with 10 elements and finally the CS3 and CS4 classes with one element each (Table 3).

For prediction we use a modular neural network, where the first module consists of a NAR network (NAR1), the second module of a modular NAR network (Modular NAR1) and the third module of a modular NAR network (Modular NAR2 Cluster).

Table 2 Clustering of GNI time series by country

Country code	Cluster ID	Country code	Cluster ID	Country code	Cluster ID	Country code	Cluster ID
AUS	CS1	FIN	CS1	MEX	CS1	CHE	CS1
AUT	CS1	GRC	CS1	NLD	CS1	TUR	CS1
BEL	CS1	HUN	CS1	NZL	CS1	FRA	CS2
CAN	CS1	ISL	CS1	NOR	CS1	DEU	CS2
CHL	CS1	IRL	CS1	POL	CS1	ITA	CS2
COL	CS1	ISR	CS1	PRT	CS1	JPN	CS2
CRI	CS1	KOR	CS1	SVK	CS1	GBR	CS2
CZE	CS1	LVA	CS1	SVN	CS1	USA	CS3
DNK	CS1	LTU	CS1	ESP	CS1		
EST	CS1	LUX	CS1	SWE	CS1		

Table 3 Clustering of population time series

Country code	Cluster ID	Country code	Cluster ID	Country code	Cluster ID	Country code	Cluster ID
AUS	CS1	GRC	CS1	NOR	CS1	KOR	CS2
AUT	CS1	HUN	CS1	PRT	CS1	MEX	CS2
BEL	CS1	ISL	CS1	SVK	CS1	POL	CS2
CAN	CS1	IRL	CS1	SVN	CS1	ESP	CS2
CHL	CS1	ISR	CS1	SWE	CS1	TUR	CS2
CRI	CS1	LVA	CS1	CHE	CS1	GBR	CS2
CZE	CS1	LTU	CS1	COL	CS2	JPN	CS3
DNK	CS1	LUX	CS1	FRA	CS2	USA	CS4
EST	CS1	NLD	CS1	DEU	CS2		
FIN	CS1	NZL	CS1	ITA	CS2		

In this case as input for each module receives a segment of the data (formed by a competitive network). Also, the final integration of results obtained by averaging the results of each module can be seen in Tables 4 and 5. They show the relative root mean squared error (RMSE).

Table 4 Prediction of GNI time series

Method	Average %RMSE	Best %RMSE	Worst %RMSE
NAR1	0.001400470	0.000707375	0.002611694
Modular NAR1	0.001421085	0.000718226	0.002698286
Modular NAR2 Cluster	0.009458815	0.001858983	0.022013062
Final prediction (integration)	0.004093457	0.001094861	0.009107681

Table 5 Prediction of population time series

Method	Average %RMSE	Best %RMSE	Worst %RMSE
NAR1	0.000180752	0.00001776	0.000423866
Modular NAR1	0.000168653	0.00005415	0.000385271
Modular NAR2 Cluster	0.000834394	0.00003572	0.004684953
Final prediction (integration)	0.000394600	0.00003588	0.001831363

5 Discussion of Results

In this work we seek to make time series prediction by using modular neural networks, with the idea of having multiple predictions that are integrated to obtain a result, which is compared with the rest of the predictions of each neural network, with the purpose of selecting the best one.

In the case of the GNI variable, the results obtained with the NAR1 and Modular NAR1 networks are very similar and slightly better than those obtained by the Modular NAR2 network. The final prediction result is better than the Modular NAR2 Cluster network, but worse than the NAR1 and Modular NAR1 networks. For the population variable, the results obtained for each network are like those presented with the GNI series.

It should be noted that for the Modular NAR2 Cluster network, the data segments formed with the competitive network represent the number of elements (countries) in each cluster. So, three clusters for the GNI series and four clusters for the population series were loaded towards a single cluster, that contained more than 60% of the elements. It was a factor for the prediction results to be similar to those obtained with the NAR1 modular network.

6 Conclusions

We have presented a model to predict population and gross national income (GNI) time series using a modular perspective in a neural network model by using both supervised and unsupervised networks in the model.

Also, we identified the advantages of using multiple monolithic and modular neural networks to predict time series, since the individual results obtained by each network can be compared against the result of the integration of all the results, and select the best of them, to improve the quality of the prediction.

For future work, we will consider using a modular neural network, where each module is a type of supervised neural network, for example: Radial Basis Function (RBF), Nonlinear Autoregressive Exogenous (NARX), Long Short-Term Memory (LSTM), among others. In addition, we could utilize metaheuristics to optimize the neural networks, as in [22–26]. Also, incorporate fuzzy logic aspects into the modular neural network aggregation process [27, 28].

References

1. E. Egrioglu, E. Bas, A new hybrid recurrent artificial neural network for time series forecasting. Neural Comput. Applic. **35**, 2855–2865 (2023). https://doi.org/10.1007/s00521-022-07753-w
2. M. Pirani, P. Thakkar, P. Jivrani, P.M. Bohara, D. Garg. A comparative analysis of ARIMA, GRU, LSTM and BiLSTM on Financial Time Series Forecasting, 2022 IEEE International

Conference on Distributed Computing and Electrical Circuits and Electronics (ICDCECE), Ballari, India, pp. 1–6 (2022). https://doi.org/10.1109/ICDCECE53908.2022.9793213

3. V. Kan, O. Alsova. Forecasting Meteorological indicators based on neural networks, 2022 IEEE International Multi-Conference on Engineering, Computer and Information Sciences (SIBIRCON), Yekaterinburg, Russian Federation, pp. 1620–1625 (2022). https://doi.org/10.1109/SIBIRCON56155.2022.10017124
4. J.L. Wu, M. Lu, C.Y. Wang. Forecasting metro rail transit passenger flow with multiple-attention deep neural networks and surrounding vehicle detection devices. Appl. Intell. (2023). https://doi.org/10.1007/s10489-023-04483-x
5. S. Xu, W. Li, Y. Zhu et al., A novel hybrid model for six main pollutant concentrations forecasting based on improved LSTM neural networks. Sci. Rep. **12**, 14434 (2022). https://doi.org/10.1038/s41598-022-17754-3
6. P.K. Sehrawat, D.K. Vishwakarma. Comparative analysis of time series models on COVID-19 predictions, 2022 International Conference on Sustainable Computing and Data Communication Systems (ICSCDS), Erode, India, pp. 710–715 2022. https://doi.org/10.1109/ICSCDS53736.2022.9760992
7. M. Ali, M.A. Syed, M. Khalid, NARX recurrent neural network based short term residential load forecasting considering the effects of multiple weather features, 2022 IEEE IAS Global Conference on Emerging Technologies (GlobConET), Arad, Romania, pp. 557–561 (2022). https://doi.org/10.1109/GlobConET53749.2022.9872509
8. M.M. Rahman, M. Shakeri, F. Khatun et al., A comprehensive study and performance analysis of deep neural network-based approaches in wind time-series forecasting. J. Reliable Intell. Environ. (2022). https://doi.org/10.1007/s40860-021-00166-x
9. F. Sohrabi, M. Reza, R. Mirabbasi, M. Tahroudi, Daily solar radiation estimation in Belleville station, Illinois, using ensemble artificial intelligence approaches. Eng. Appl. Artif. Intell. **120**, 105839 (2023). https://doi.org/10.1016/j.engappai.2023.105839
10. J. Siłka, M. Wieczorek, M. Woźniak, Recurrent neural network model for high-speed train vibration prediction from time series. Neural Comput. Applic. **34**, 13305–13318 (2022). https://doi.org/10.1007/s00521-022-06949-4
11. X. Huang, S. Yoo, A deep neural network for multivariate time series clustering with result interpretation, 2021 International Joint Conference on Neural Networks (IJCNN), Shenzhen, China, 1–8 (2021). https://doi.org/10.1109/IJCNN52387.2021.9533427
12. H. Wang et al., Electric vehicle charging load clustering and load forecasting based on long short term memory neural network, 2022 IEEE 5th International Electrical and Energy Conference (CIEEC), Nangjing, China, pp. 3196-3200 (2022). https://doi.org/10.1109/CIEEC54735.2022.9846570
13. J. Yao, B. Lu, J. Zhang, Multi-step-ahead tool state monitoring using clustering feature-based recurrent fuzzy neural networks. IEEE Access **9**, 113443–113453 (2021). https://doi.org/10.1109/ACCESS.2021.3104668
14. V. Bhardwaj, P. Bhavsar, D. Patnaik, Forecasting GDP per capita of OECD countries using machine learning and deep learning models, 2022 Interdisciplinary Research in Technology and Management (IRTM), Kolkata, India, pp. 1-6 (2022). https://doi.org/10.1109/IRTM54583.2022.9791714
15. The World Bank Data: Population, total (2022), June 06. Retrieved from: https://data.worldbank.org/indicator/SP.POP.TOTL
16. The World Bank Data: GNI per capita, 2022, June 06. Retrieved from: https://data.worldbank.org/indicator/NY.GNP.MKTP.CD
17. J.S. Jang, E. Mizutani, Unsupervised learning and other neural networks. In: Neuro-fuzzy and soft computing: a computational approach to learning and machine Intelligence, Prentice Hall, Upper Saddle River, NJ 07458, pp. 301–310 (1997)
18. E. Méndez, O. Lugo, P. Melin, A competitive modular neural network for long-term time series forecasting. In: P. Melin, O. Castillo, J. Kacprzyk (eds.) Nature-Inspired Design of Hybrid Intelligent Systems, Studies in Computational Intelligence, vol 667. Springer, pp. 243–254 (2012). https://doi.org/10.1007/978-3-319-47054-2_16

19. M. Pulido, P. Melin, Comparison of genetic algorithm and particle swarm optimization of ensemble neural networks for complex time series prediction. In: Melin P., Castillo O., Kacprzyk J. (eds) Recent Advances of Hybrid Intelligent Systems Based on Soft Computing. Studies in Computational Intelligence, vol. 915. Springer, Cham., pp. 51–77 (2021). https://doi.org/10.1007/978-3-030-58728-4_3imagenes
20. M. Pulido, P. Melin, Ensemble recurrent neural networks and their optimization by particle swarm for complex time series prediction. In: O. Castillo, P. Melin (eds.), New Perspectives on Hybrid Intelligent System Design based on Fuzzy Logic, Neural Networks and Metaheuristics. Studies in Computational Intelligence, vol 1050. Springer, Cham., pp. 47–61 (2022). https://doi-org.pbidi.unam.mx:2443/10.1007/978-3-031-08266-5_4
21. J.C. Mónica, P. Melin, D. Sánchez, Genetic optimization of ensemble neural network architectures for prediction of COVID-19 confirmed and death cases. In: O. Castillo, P. Melin (eds.) Fuzzy Logic Hybrid Extensions of Neural and Optimization Algorithms: Theory and Applications. Studies in Computational Intelligence, vol 940. Springer, Cham, pp. 85–98 (2021). https://doi.org/10.1007/978-3-030-68776-2_5
22. B. González, F. Valdez, P. Melin, G. Prado-Arechiga, Fuzzy logic in the gravitational search algorithm for the optimization of modular neural networks in pattern recognition. Expert Syst. Appl. **42**(14), 5839–5847 (2015)
23. F. Valdez, J.C. Vazquez, P. Melin, O. Castillo, Comparative study of the use of fuzzy logic in improving particle swarm optimization variants for mathematical functions using co-evolution. Appl. Soft Comput. **52**, 1070–1083 (2017)
24. D. Sanchez, P. Melin, O. Castillo, A grey wolf optimizer for modular granular neural networks for human recognition. Comput. Intell. Neurosci. **2017**(2017). https://doi.org/10.1155/2017/4180510
25. O. Castillo, E. Lizzarraga, J. Soria, P. Melin, F. Valdez, New approach using ant colony optimization with ant set partition for fuzzy control design applied to the ball and beam system. Inf. Sci. **294**, 203–215 (2015)
26. L. Amador-Angulo, O. Mendoza, J.R. Castro, A. Rodriguez-Diaz, P. Melin, O. Castillo, Fuzzy sets in dynamic adaptation of parameters of a bee colony optimization for controlling the trajectory of an autonomous mobile robot. Sensors **16**(9), 1458 (2016)
27. E. Ontiveros, P. Melin, O. Castillo, Comparative study of interval type-2 and general type-2 fuzzy systems in medical diagnosis. Inf. Sci. **525**, 37–53 (2020)
28. J.E. Moreno, M.A. Sanchez, O. Mendoza, A. Rodriguez-Diaz, O. Castillo, P. Melin, J.R. Castro, Design of an interval Type-2 fuzzy model with justifiable uncertainty. Inf. Sci. **513**, 206–221 (2020)

Aggregation in Ensemble Neural Models with Type-3 and Intuitionistic Fuzzy Logic

Patricia Melin and Oscar Castillo

1 Introduction

Fuzzy logic has become very important in different disciplines of study, one of the areas in which we focus for this work is the time series prediction area, it has been shown in the literature that the use of fuzzy logic helps to improve results in many problems [1, 2]. Type-1 evolved to type-2 fuzzy systems with the works by Mendel in 2001 [3]. Initially, interval type-2 fuzzy systems were studied and applied to several problems [3]. These systems were applied to many problems in areas such as: robotics, control, diagnosis and others [4, 5]. Simulation and experimental results show that interval type-2 outperforms type-1 fuzzy systems in situations with higher levels of noise, dynamic environments or highly nonlinear problems [6–8]. Later, general type-2 fuzzy systems were considered to manage higher levels of uncertainty, and good results have been achieved in several areas of application [9–11]. Recently, type-3 has been able to solve complex problems. For this reason, in this paper we are putting forward the basic constructs of type-3 fuzzy systems by extending the ideas of type-2 [12–14].

Few year ago, the propagation of COVID-19 was critical. In particular, in the case of Europe several countries, like Italy, Spain, and France have been hit hard with the spread of the COVID-19 virus, having a significant number of confirmed cases and deaths [15–20]. In the case of the American continent, United States, Canada and Brazil have also suffer a significant number of cases due to the rapid spread of COVID-19 [21–23]. There are also several recent works on predicting and modeling COVID-19 behavior in space and time [24, 25].

The contribution is the type-3 fuzzy intuitionistic idea to enhance prediction in ensembles of neural models. This has not been previously presented, and it is

P. Melin (✉) · O. Castillo
Tijuana Institute of Technology, TecNM, Tijuana, BC, Mexico
e-mail: pmelin@tectijuana.mx

P. Melin and O. Castillo (eds.), *New Directions on Hybrid Intelligent Systems Based on Neural Networks, Fuzzy Logic, and Optimization Algorithms*, Studies in Computational Intelligence 1146, https://doi.org/10.1007/978-3-031-53713-4_7

now shown that type-3 intuitionistic can surpass other methods. We view this as an important contribution to the frontiers in fuzzy theory.

The article structured as: Sect. 2 defines type-3 and intuitionistic fuzzy sets, Sect. 3 puts forward the proposal, Sect. 4 exhibits preliminary results, and Sect. 5 offers the conclusions.

2 Type-3 and Intuitionistic

Type-3 is as an extension of type-2. We first offer type-3 fuzzy terminology and then the combination with intuitionistic theory.

2.1 Type-3 Fuzzy Sets

The type-3 set is postulated as:

Definition 1 A type-3 fuzzy set (T3 FS) [27–29], denoted by $A^{(3)}$, is represented by a membership function (MF) of $A^{(3)}$, in the Cartesian product $X \times [0, 1] \times [0, 1]$ in $[0, 1]$, where X is the universe of the primary variable of $A^{(3)}$, x. The MF of $\mu_{A^{(3)}}$ is postulated by $\mu_{A^{(3)}}(x, u, v)$ (or $\mu_{A^{(3)}}$) and it is called a type-3 MF (T3 MF) of the T3 FS:

$$\mu_{A^{(3)}} : X \times [0, 1] \times [0, 1] \rightarrow [0, 1]$$

$$A^{(3)} = \{(x, u(x), v(x, u), \mu_{A^{(3)}}(x, u, v)) | x \in X, u \in U \subseteq [0, 1], v \in V \subseteq [0, 1]\} \tag{1}$$

where U is the universe for the secondary variable u and V is the universe for tertiary variable v. If the tertiary MF is uniformly equal to 1 then we have an Interval type-3 fuzzy set (IT3 FS) with interval type-3 MF (IT3MF).

Figure 1 depicts an IT3MF $\tilde{\mu}(x, u)$, where $\underline{\mu}(x, u)$ is the lower MF and $\overline{\mu}(x, u)$ is the upper MF. The secondary T1 MFs in x' are $\underline{f}_{x'}(u)$ and $\overline{f}_{x'}(u)$.

2.2 Type-3 Intuitionistic

Now we extend type-3 to the intuitionistic realm. Basically, we now introduce the tertiary MF to represent the membership over the secondary MF [15]. The idea is

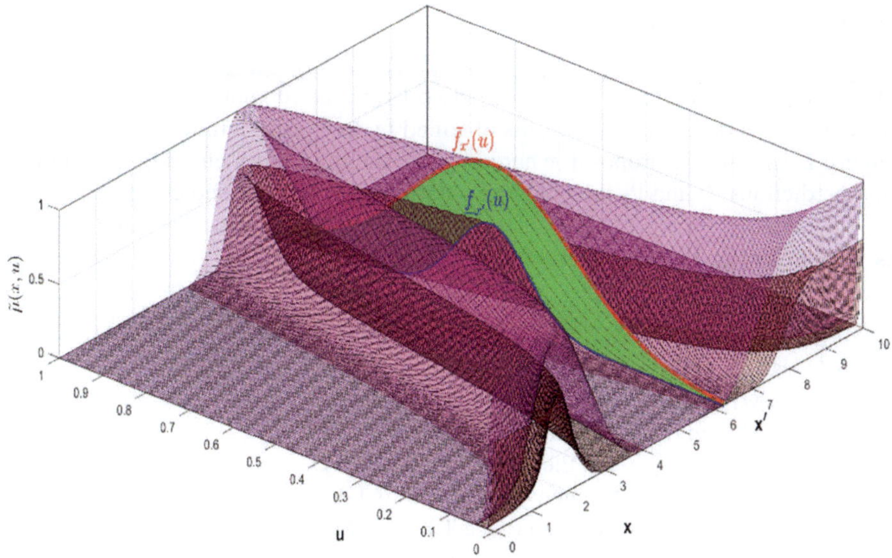

Fig. 1 Fuzzy set with IT3 MF $\widetilde{\mu}\ (x, u)$

providing a better capability for handling uncertainty. A type-3 Intuitionistic FS is postulated as:

$$A = \{\langle x, \mu_{A1}(x, u, v), n_{A1}(x, u, v)\rangle \,|\, x \in X, \ u \in J_x, \ v \in H_x\}, \quad (2)$$

where $\mu_{AI}(x, u, v)$ and $v_{AI}(x, u, v)$ are the membership and non-membership degrees of $x \in X$, J_x is the secondary domain, H_x is the tertiary domain, and satisfies

$$0 \leq \mu_{A1}(x, u, v) + v_{A1}(x, u, v) \leq 1 \quad (3)$$

We can utilize the following expression to approximate the results of type-3 intuitionistic:

$$T3IFIS = (1 - \pi)T3FIS\mu + \pi T3FISv \quad (4)$$

Evidently, there are a plethora of theoretical issues to consider as future work.

3 Proposed Method

We consider an ensemble of two neural networks and then combine their outputs using a weighted average with the weights obtained by the type-3 intuitionistic approach. Figure 2 exhibits the proposal, where we note the time series entering the modules and predictions P_1 and P_2 are obtained with training errors e_1 and e_2, respectively.

For example, for a genetic algorithm, P_1 could be the crossover rate and P_2 could be mutation rate. The fuzzy rules for aggregation are:

1. If (e_1 is S) and (e_2 is S), then (w_1 is H)(w_2 is H).
2. If (e_1 is S) and (e_2 is M), then (w_1 is H)(w_2 is M).
3. If (e_1 is S) and (e_2 is H), then (w_1 is H)(w_2 is L).
4. If (e_1 is M) and (e_2 is S), then (w_1 is M)(w_2 is H).
5. If (e_1 is M) and (e_2 is M), then (w_1 is M)(w_2 is M).
6. If (e_1 is M) and (e_2 is H), then (w_1 is M)(w_2 is L).
7. If (e_1 is H) and (e_2 is S), then (w_1 is L)(w_2 is H).
8. If (e_1 is H) and (e_2 is M), then (w_1 is L)(w_2 is M).
9. If (e_1 is H) and (e_2 is H), then (w_1 is L)(w_2 is L).

In this case, S means small, M mean medium and H means high. The fuzzy rule design was based on neural network training knowledge. This knowledge was utilized for expressing the rules. The type-3 system (in Fig. 3) has as inputs the errors of the networks, e_1 and e_2, respectively. After defuzzification, the Type-3 intuitionistic system produces the weights (w_1 and w_2) to obtain a final prediction P given by:

$$P = \frac{w_1 P_1 + w_2 P_2}{w_1 + w_2} \tag{5}$$

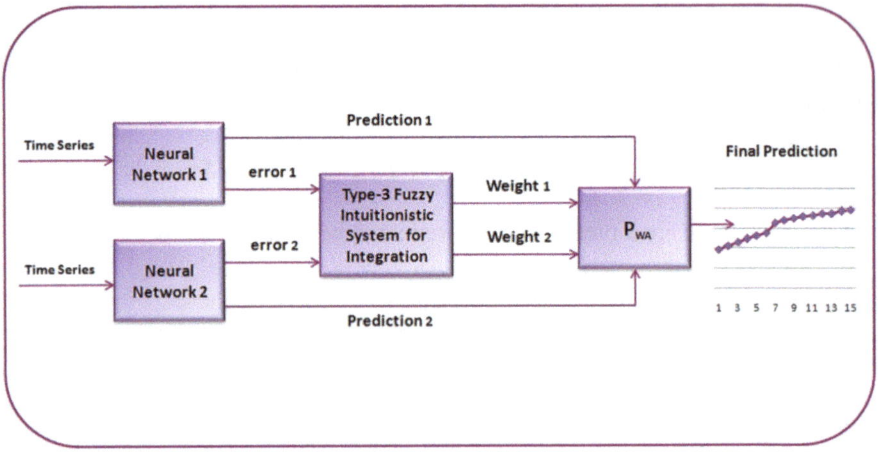

Fig. 2 Structure of the type-3 intuitionistic aggregation in ensembles

Table 1 summarizes the MF parameterization, which was done by trial and error. Table 1 exhibits centers and deviations of the MFs.

Lower scale (λ) and lower lag (ℓ) parameters were found to be 0.8 and 0.2 for the inputs, respectively. For the outputs, they were found to be 0.9 and 0.6, respectively.

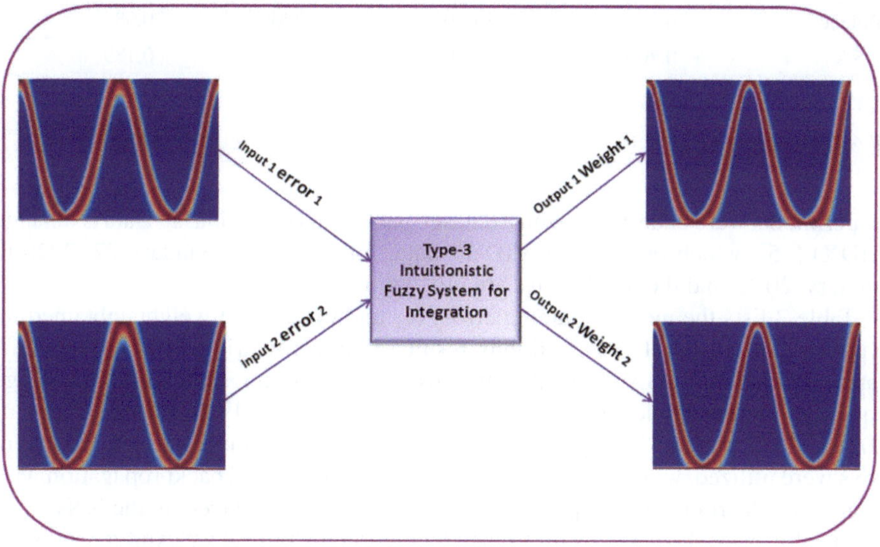

Fig. 3 Type-3 intuitionistic system to obtain the weights

Table 1 Parameterization of the Gaussian MFs

Variable	MF	σ	m
error 1	S	0.10	0.00
error 1	M	0.12	0.50
error 1	H	0.10	1.00
error 2	S	0.10	0.00
error 2	M	0.12	0.50
error 2	H	0.10	1.00
weight 1	L	0.10	0.00
weight 1	M	0.11	0.50
weight 1	H	0.10	1.00
weight 2	L	0.10	0.00
weight 2	M	0.11	0.50
weight 2	H	0.10	1.00

Table 2 Error data and weights produced by type-3 intuitionistic

Country	e_1	e_2	W_1	W_2
France	1.000	0.820	0.165	0.186
Germany	0.956	0.110	0.094	0.913
Japan	1.000	0.860	0.113	0.118
Poland	0.0958	1.000	0.920	0.087
USA	0.700	1.000	0.496	0.139

4 Results

Experiments were undertaken with a dataset used from Humanitarian Data Exchange (HDX) [15], which includes COVID-19 data, for cases from January 22, 2020 to January, 2022, and the last 15 days are utilized for testing.

Table 2 lists the module training errors (e_1 and e_2) and the weights obtained by type-3 intuitionistic. Table 2 exhibits results of five countries: France, Germany, Japan, Poland and USA. Table 2 illustrates the prediction results with the average equation. The ensemble modules were trained with COVID-19 data from January of 2020 to 2022, and last 15 days are utilized for comparing with real values. Recurrent NNs were utilized, with three delays, 300 training epochs, and backpropagation with momentum learning and adaptive learning rate. There are 3 layers in the NNs.

In Table 3 we show results for France prediction. In Table 4 we exhibit results for Germany. In Table 5 we list results for Japan. In addition, we show in Table 6 results for Poland. Lastly, we summarize in Table 7 the USA prediction.

Based on Table 3, results for France are close to the real values, validating the proposal.

Germany prediction results are very good based on Table 4, where the forecasted and real values are very near.

Japan prediction results are good according to Table 5, where forecasted and real values are near.

According to Table 6 results for Poland are close to the real values. Finally, in Table 7 we can also note predicted and real values for USA.

5 Conclusions

In this work an aggregation approach for ensemble models has been outlined. In this case a type-3 intuitionistic fuzzy system is employed for weight estimation utilized in the weighted average. Uncertainty in aggregation is modeled with type-3 and intuitionistic fuzzy. The proposal was verified with real time series of five countries with good results. As future work we plan to test the proposal in other applications, like the ones presented in [30–33]. In addition, we plan to optimize the fuzzy system design with metaheuristics, such as in [34–39].

Table 3 Comparison of prediction results with the proposal and real values for France

P_1	P_2	P_{WA}	P_{Real}
6,605,410.19	7,008,247.76	6,819,020.56	7,075,244
6,614,139.24	7,017,080.51	6,827,804.6	7,079,005
6,622,399.29	7,019,209.37	6,832,813.5	7,093,651
6,627,109.31	7,030,190.08	6,840,848.64	7,106,147
6,638,009.1	7,038,745.08	6,850,505.07	7,109,125
6,646,165.48	7,040,257.54	6,855,138.42	7,128,903
6,651,161.41	7,055,100.1	6,865,355.67	7,149,118
6,666,269.19	7,069,013.04	6,879,829.87	7,168,026
6,681,418.33	7,081,862.44	6,893,759.54	7,188,721
6,696,029.19	7,096,088.87	6,908,166.55	7,211,399
6,711,976.63	7,111,599.88	6,923,882.57	7,231,148
6,728,648.35	7,124,729.52	6,938,676.05	7,235,966
6,741,361.56	7,126,988.88	6,945,845.95	7,266,361
6,748,887.39	7,149,216.18	6,961,167.45	7,296,757
6,771,633.28	7,169,428.36	6,982,569.8	7,330,086

Table 4 Comparison of prediction results with the proposal and real values for Germany

P_1	P_2	P_{WA}	P_{Real}
4,874,657.83	8,186,604.82	7,874,981.02	7,866,784
4,877,993.44	8,284,832.8	7,964,280.49	7,943,959
4,880,960.91	8,372,079.56	8,043,597.35	7,988,210
4,882,089.4	8,414,449.58	8,082,086.92	8,021,339
4,882,788.23	8,444,313.88	8,109,207.02	8,104,157
4,886,304.01	8,546,418.12	8,202,034.98	8,222,262
4,891,326.88	8,696,895.89	8,338,826.79	8,361,262
4,896,263.17	8,864,841.73	8,491,434.92	8,502,132
4,900,498.6	9,027,704.27	8,639,372.11	8,635,461
4,903,941.72	9,175,951.89	8,773,994.95	8,716,804
4,905,466.24	9,255,829.58	8,846,500.33	8,773,030
4,906,195.81	9,304,413.44	8,890,581.54	8,909,503
4,909,695.04	9,465,430.18	9,036,777.33	9,088,672
4,914,154.61	9,684,141.65	9,235,329.67	9,317,280
4,918,690.59	9,948,751.83	9,475,469.25	9,477,603

Table 5 Comparison of prediction results with the proposal and real values for Japan

P_1	P_2	P_{WA}	P_{Real}
1,421,031.66	1,168,372.61	1,292,086.17	1,209,082
1,435,781.23	1,191,360.06	1,311,039.97	1,234,423
1,449,848.17	1,213,401.89	1,329,176.93	1,260,415
1,462,806.61	1,235,123.08	1,346,607.47	1,286,077
1,474,669.31	1,255,662.63	1,362,898.43	1,308,422
1,484,070.3	1,271,603.87	1,375,637.26	1,325,340
1,517,540.24	1,340,270.56	1,427,069.99	1,420,783
1,525,625.61	1,357,837.38	1,439,994.25	1,443,642
1,532,582.42	1,373,612.4	1,451,451.47	1,462,979
1,537,962.41	1,385,600.95	1,460,204.16	1,476,653
1,541,193.63	1,392,303.61	1,465,207.05	1,494,372
1,545,591.19	1,406,220.89	1,474,463.04	1,514,400
1,545,591.19	1,406,220.89	1,474,463.04	1,514,400
1,551,355.7	1,421,248.78	1,484,955.15	1,532,616
1,556,035.67	1,432,831.96	1,493,158.2	1,549,337

Table 6 Comparison of prediction results with the proposal and real values for Poland

P_1	P_2	P_{WA}	P_{Real}
3,995,819.87	3,621,223.9	3,963,176.88	3,881,349
4,003,077.22	3,637,991.57	3,971,262.98	3,903,445
4,008,232.79	3,647,846.54	3,976,828.06	3,923,472
4,013,534.37	3,659,477.29	3,982,681.17	3,942,864
4,018,723.82	3,671,368.56	3,988,454.63	3,958,840
4,022,446.88	3,679,114.97	3,992,528.3	3,968,450
4,024,370.78	3,682,395.61	3,994,570.43	3,982,257
4,028,881.2	3,694,834.14	3,999,771.72	4,000,270
4,033,453.12	3,706,100.13	4,004,926.97	4,017,420
4,037,113.47	3,714,046.58	4,008,960.82	4,032,796
4,040,580.05	3,722,064.06	4,012,823.97	4,043,585
4,042,683.73	3,726,273.81	4,015,111.18	4,049,838
4,043,955.25	3,728,860.68	4,016,497.32	4,054,865
4,045,314.96	3,732,474.08	4,018,053.42	4,064,715
4,048,172.85	3,740,829.22	4,021,390.35	4,080,282

Table 7 Comparison of prediction results with the proposal and real values for USA

P1	P_2	P_{WA}	P_{Real}
61,326,847	60,718,991.9	61,193,732.4	64,210,668
62,011,425.7	61,319,440.6	61,859,887.4	65,069,619
62,675,140.7	61,930,734.2	62,512,122.6	65,463,200
62,750,945.4	62,101,842.9	62,608,798	65,930,556
62,982,269.4	62,412,455.5	62,857,485.4	66,590,148
63,622,278.3	62,921,873.6	63,468,896.2	67,693,339
64,773,747.8	63,794,454.8	64,559,291.7	68,684,431
65,636,754.3	64,470,771.5	65,381,414.9	69,388,781
65,982,382.4	64,869,656.6	65,738,705.8	70,206,083
66,548,711.7	65,413,568.4	66,300,125.8	70,490,987
66,527,725.9	65,492,563	66,301,034.9	70,845,794
66,625,713.4	65,702,223.5	66,423,477.7	71,741,698
67,597,051.5	66,400,650.5	67,335,050.8	72,257,016
67,982,681.7	66,685,204.4	67,698,546.2	72,910,136
68,386,962.5	67,101,527.1	68,105,464.1	73,427,335

References

1. L.A. Zadeh, Knowledge representation in Fuzzy Logic. IEEE Trans. Knowl. Data Eng. **1**, 89–100 (1989)
2. L.A. Zadeh, Fuzzy logic. Computer **1**(4), 83–93 (1998)
3. J.M. Mendel, *Uncertain rule-based fuzzy logic systems: introduction and new directions* (Prentice-Hall, Upper-Saddle River, 2001)
4. J.M. Mendel, Uncertain rule-based fuzzy logic systems: introduction and new directions, Second Edition, Springer (2017)
5. N.N. Karnik, J.M. Mendel, Operations on type-2 fuzzy sets. Fuzzy Sets Syst. **122**, 327–348 (2001)
6. J.E. Moreno et al., Design of an interval Type-2 fuzzy model with justifiable uncertainty. Inf. Sci. **513**, 206–221 (2020)
7. J.M. Mendel, H. Hagras, W.-W. Tan, W.W. Melek, H. Ying, *Introduction to type-2 fuzzy logic control* (Wiley and IEEE Press, Hoboken, NJ, 2014)
8. F. Olivas, F. Valdez, O. Castillo, P. Melin, Dynamic parameter adaptation in particle swarm optimization using interval type-2 fuzzy logic. Soft. Comput. **20**(3), 1057–1070 (2016)
9. A. Sakalli, T. Kumbasar, J.M. Mendel, Towards systematic design of general type-2 fuzzy logic controllers: analysis, interpretation, and tuning. IEEE Trans. Fuzzy Syst. **29**(2), 226–239 (2021)
10. E. Ontiveros, P. Melin, O. Castillo, High order α-planes integration: a new approach to computational cost reduction of general type-2 fuzzy systems. Eng. Appl. Artif. Intell. **74**, 186–197 (2018)
11. O. Castillo, L. Amador-Angulo, A generalized type-2 fuzzy logic approach for dynamic parameter adaptation in bee colony optimization applied to fuzzy controller design. Inf. Sci. **460**, 476–496 (2018)
12. Y. Cao, A. Raise, A. Mohammadzadeh, S.S. Rathinasamy, S. Band, A. Mosavi, Deep learned recurrent type-3 fuzzy system: application for renewable energy modeling / prediction. Energy Rep. **7**, 8115–8127 (2021)

13. A. Mohammadzadeh, O. Castillo, S.S. Band, A. Mosavi, A novel fractional-order multiple-model type-3 fuzzy control for nonlinear systems with unmodeled dynamics. Int. J. Fuzzy Syst. **23**, 1633–1651 (2021)
14. S.N. Qasem, A. Ahmadian, A. Mohammadzadeh, S. Rathinasamy, B. Pahlevanzadeh, A type-3 logic fuzzy system: optimized by a correntropy based Kalman filter with adaptive fuzzy kernel size. Inform. Sci. **572**, 424–443 (2021)
15. The Humanitarian Data Exchange (HDX)," [Online]. Available: https://data.humdata.org/dataset/novel-coronavirus-2019-ncov-cases. [Accessed 31 03 2020]
16. M.A. Shereen, S. Khan, A. Kazmi, N. Bashir, R. Siddique, COVID-19 infection: origin, transmission, and characteristics of human coronaviruses. J. Adv. Res. **24**, 91–98 (2020)
17. C. Sohrabi, Z. Alsafi, N. O'Neill, M. Khan, A. Kerwan, A. Al-Jabir ... R. Agha, World health organization declares global emergency: a review of the 2019 novel coronavirus (COVID-19). Intern. J. Surg. **76**, 71-76
18. I.D. Apostolopoulos, T. Bessiana, Covid-19: automatic detection from X-Ray images utilizing Transfer Learning with Convolutional Neural Networks (2020). *arXiv preprint* arXiv:2003.11617
19. S.A. Sarkodie, PA. Owusu, Investigating the cases of novel coronavirus disease (COVID-19) in China Using Dynamic Statistical Techniques. *Available at SSRN 3559456* (2020)
20. B.R. Beck, B. Shin, Y. Choi, S. Park, K. Kang, Predicting commercially available antiviral drugs that may act on the novel coronavirus (SARS-CoV-2) through a drug-target interaction deep learning model. Comput. Struct. Biotechnol. J. **18**, 784–790 (2020)
21. L. Zhong, L. Mu, J. Li, J. Wang, Z. Yin, D. Liu, Early prediction of the 2019 novel coronavirus outbreak in the Mainland China based on simple mathematical model. IEEE Access. **8**, 51761–51769 (2020)
22. M.N. Kamel Boulos, E.M. Geraghty, Geographical tracking and mapping of coronavirus disease COVID-19/severe acute respiratory syndrome coronavirus 2 (SARS-CoV-2) epidemic and associated events around the world: how 21st century GIS technologies are supporting the global fight against outbreaks and epidemics. Int. J. Health Geogr.**19**(1)**,** 8, 1–12 (2020)
23. P. Gao, H. Zhang, Z. Wu, J. Wang, Visualising the expansion and spread of coronavirus disease 2019 by cartograms. Environ Plan A **52**(4), 698–701 (2020)
24. A.S.R.S. Rao, J.A. Vazquez, Identification of COVID-19 can be quicker through artificial intelligence framework using a mobile phone-based survey in the populations when Cities/Towns are under quarantine. Infect. Control Hosp. Epidemiol. **41**(7), 826–830 (2020)
25. P. Melin, J.C. Monica, D. Sanchez, O. Castillo, (2020) Analysis of spatial spread relationships of coronavirus (COVID-19) pandemic in the world using self organizing maps. Chaos Solitons Fractals **138**(109917), 1–7 (2020)
26. P. Melin, J.C. Monica, D. Sanchez, O. Castillo, Multiple ensemble neural network models with fuzzy response aggregation for predicting COVID-19 time series: the case of Mexico. Healthcare **8**, 181 (2020)
27. J.T. Rickard, J. Aisbett, G. Gibbon, Fuzzy subsethood for fuzzy sets of type-2 and generalized type-n. IEEE Trans. Fuzzy Syst. **17**(1), 50–60 (2009)
28. A. Mohammadzadeh, M.H. Sabzalian, W. Zhang, An interval type-3 fuzzy system and a new online fractional-order learning algorithm: theory and practice. IEEE Trans. Fuzzy Syst. **28**(9), 1940–1950 (2020)
29. Z. Liu, A. Mohammadzadeh, H. Turabieh, M. Mafarja, S.S. Band, A. Mosavi, A new online learned interval type-3 fuzzy control system for solar energy management systems. IEEE Access **9**, 10498–10508 (2021)
30. L. Cervantes, O. Castillo, Type-2 fuzzy logic aggregation of multiple fuzzy controllers for airplane flight control. Inf. Sci. **324**, 247–256 (2015)
31. P. Melin, O. Castillo, An intelligent hybrid approach for industrial quality control combining neural networks, fuzzy logic and fractal theory. Inf. Sci. **177**, 1543–1557 (2007)
32. O. Castillo, J.R. Castro, P. Melin, A. Rodriguez-Diaz, Application of interval type-2 fuzzy neural networks in non-linear identification and time series prediction. Soft. Comput. **18**(6), 1213–1224 (2014)

33. E. Rubio, O. Castillo, F. Valdez, P. Melin, C.I. Gonzalez, G. Martinez, An extension of the fuzzy possibilistic clustering algorithm using type-2 fuzzy logic techniques. Adv. Fuzzy Syst. **2017**(7094046), 23 (2017)
34. F. Valdez, P. Melin, O. Castillo, Evolutionary method combining particle swarm optimization and genetic algorithms using fuzzy logic for decision making, in: IEEE International Conference on Fuzzy Systems, pp. 2114–2119 (2009)
35. F. Valdez, J.C. Vazquez, P. Melin, O. Castillo, Comparative study of the use of fuzzy logic in improving particle swarm optimization variants for mathematical functions using co-evolution. Appl. Soft Comput. Comput. **52**, 1070–1083 (2017)
36. D. Sanchez, P. Melin, O. Castillo, A grey wolf optimizer for modular granular neural networks for human recognition. Comput. Intell. Neurosci. **2017**(2017). https://doi.org/10.1155/2017/4180510
37. O. Castillo, E. Lizzarraga, J. Soria, P. Melin, F. Valdez, New approach using ant colony optimization with ant set partition for fuzzy control design applied to the ball and beam system. Inf. Sci. **294**, 203–215 (2015)
38. L. Amador-Angulo, O. Mendoza, J.R. Castro, A. Rodriguez-Diaz, P. Melin, O. Castillo, Fuzzy sets in dynamic adaptation of parameters of a bee colony optimization for controlling the trajectory of an autonomous mobile robot. Sensors **16**(9), 1458 (2016)
39. B. González, F. Valdez, P. Melin, G. Prado-Arechiga, Fuzzy logic in the gravitational search algorithm for the optimization of modular neural networks in pattern recognition. Expert Syst. Appl. **42**(14), 5839–5847 (2015)

Radial Basis Function Neural Network for Time Series Prediction: The Case of Yen Currency

Martha Pulido and Patricia Melin

1 Introduction

The Radial Basis Function Network (RBFN) was proposed by Broomhead and Lowe in 1988 and has a three-layer fixed topology [1]. The layer input is responsible for receiving and distributing data from the outside. The hidden layer is activated by Radial Basis Functions (RBFs), thus generating responses with non-linear values from the input layer. The nodes in this layer exhibit a local behavior, since these are activated when the elements of the input layer are within their neighborhood, that is, they are close to the RBF centers. Finally, the layer of output is driven by continuous linear functions, so a linear combination of the outputs of the hidden layer nodes is achieved. This type of nonlinear classifier has the advantage of being very simple in terms of its architecture, unlike other types of neural networks, such as the Feed forward Neural Network or the Recurrent Neural Network [2, 5], which can have more than three layers. Another important advantage of this type of network is that it can generate hyperspheres of decisions by using RBF on the hidden layer nodes, so it is able to classify sets of non-linearly separable data [6, 12].

Radial-based neuronal networks are the ones that best suit our objective, because they are multilayer networks with forward connections, which are characterized by the fact that they are made up of a single hidden layer and each neuron in this layer has a local character., in the sense that each hidden neuron in the network it is activated in a different region of the input pattern space [13, 32]. This local character is given by the use of the so-called radial basis functions, generally the Gaussian function, as

M. Pulido · P. Melin (✉)
Tijuana Institute of Technology, TecNM, Tijuana, Baja California, Mexico
e-mail: pmelin@tectijuana.mx

M. Pulido
e-mail: martha.pulido@tectijuana.edu.mx

P. Melin and O. Castillo (eds.), *New Directions on Hybrid Intelligent Systems Based on Neural Networks, Fuzzy Logic, and Optimization Algorithms*, Studies in Computational Intelligence 1146, https://doi.org/10.1007/978-3-031-53713-4_8

activation functions. Output layer neurons radially based networks simply perform a linear combination of hidden neuron firings [20, 30].

This paper deals with the development of the ensemble radial basis network (ERBN). This design is employed for the prediction of the time series, and the structure consists of finding the number of neurons (NN), the number of the hidden layer (NL) of the (ERBN). The integration of network responses network was carried out through type-1 Fuzzy System (T1FS) and Type-2 Fuzzy System (IT2FS) and as result, we get the Prediction. The systems mentioned have three inputs which, are Pd1, Pd2, and Pd3 and one output is named Pd the range is 0 to 1 and were defined as two MFs "Lw", "Hgh", as linguistic variables are of Mamdani (FIS), as presented in Fig. 4.

This paper consists of 4 sections: In Sect. 2, the problem to solve and solution with proposed method is mentioned in detail, Sect. 3 describes in detail the results obtained and Sect. 4 describes the final conclusions reached with the implementation of this method.

2 Proposed Method

In this section, the Radial Basis Function Network (RBFN) model is presented where we have historical data, then this data enters the modules of the ERBFN, and then responses from the ERBFN are integrated with T1FS and IT2FS and therefore we obtain the prediction, in this case for: Yen/Dollar and Yen/Mexican pesos time series.

Figure 1 describes the scheme of each of the steps of the paper which is explained below.

Figure 2 presents the graph of the data Yen/Dollar [31], for which we employ 1200 data that belongs to the period from 13/06/2017 to 20/01/2022. In this case 30% of the data was used for testing and 70% for training of RBFN.

Figure 3 presents the plot of the Yen/Mexican Pesos data [312], for which we employ 1200 data points that belong to the period from 16/06/2017 to 20/01/2022. In this case 30% was employed for testing and 70% for training of RBFN.

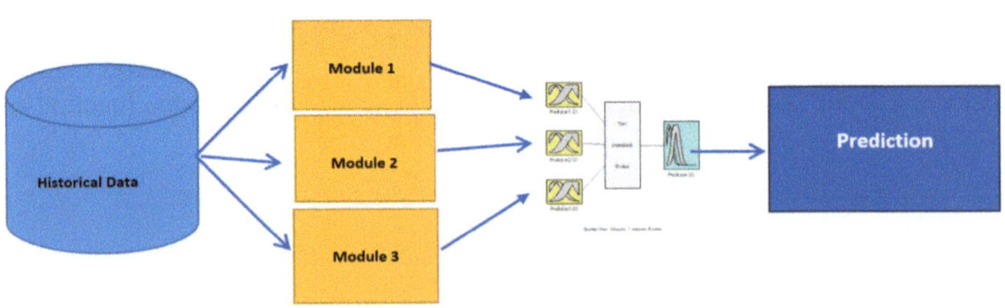

Fig. 1 Scheme of the ERBFN

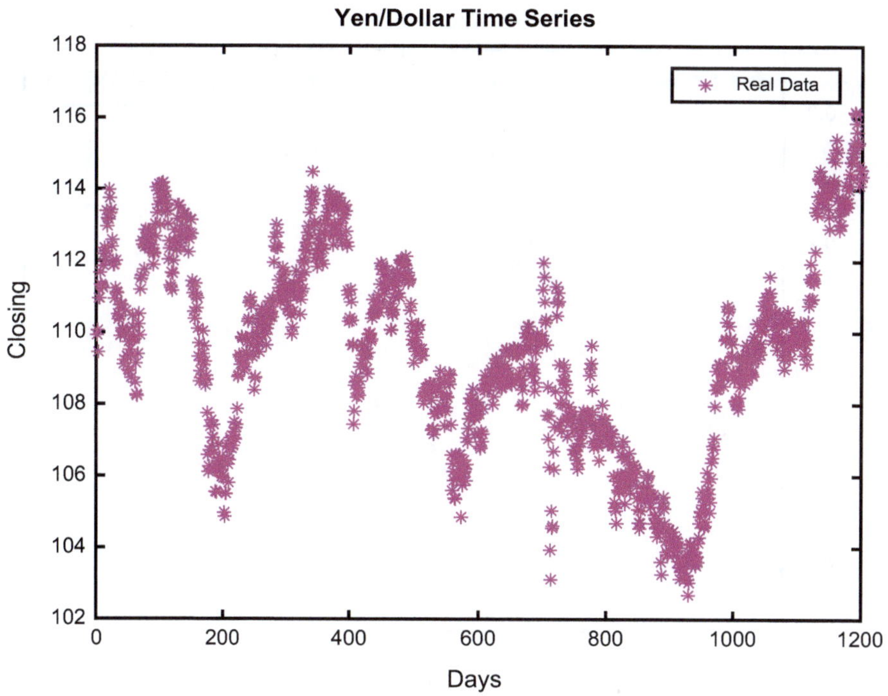

Fig. 2 Yen/Dollar data

The fuzzy system uses three inputs Pd1, Pd2 and Pd3 and the output is called Pd, where parameters are from 0 to 1.4 and linguistic values are denominated "Lw", "Hgh", as presented in Figs. 4 and 5.

The following equation shows how the total results of the FS is calculated:

$$y = \frac{\sum_{i=1x_i}^{n} u(x_i)}{\sum_{i=1}^{n} u(x_i)_i} \tag{2}$$

where u represent MFs, and x_i corresponds to the input data.

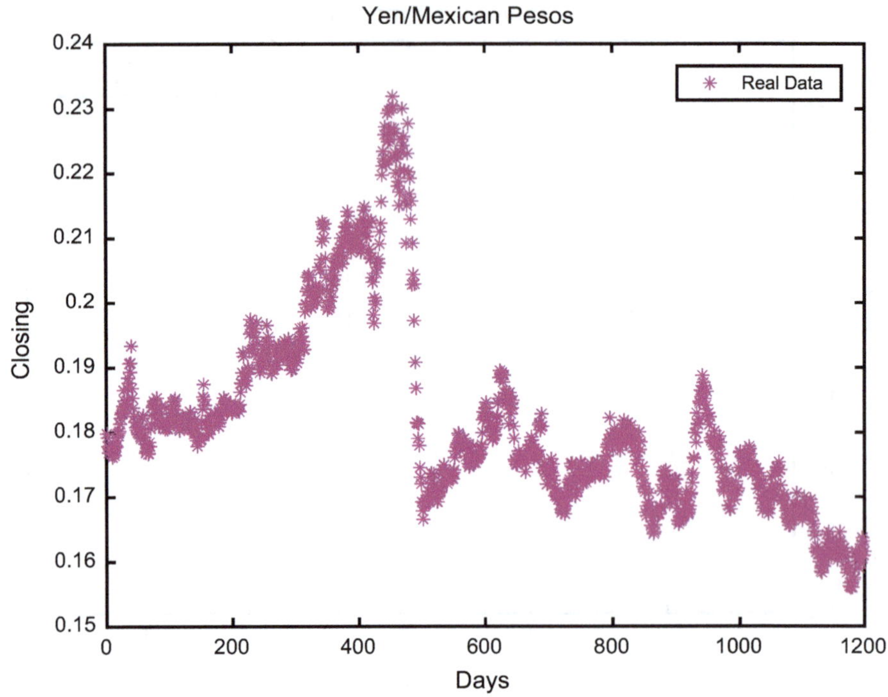

Fig. 3 Yen/Mexican pesos data

3 Simulation Results

In this part, a general description of the experiments is shown and the results of the present model are illustrated, in this case, we the radial function neural network and T1FS, IT2FS integration and 10 experiments were carried out, and the time series of the Yen/Dollar and Yen/Mexican pesos.

Table 1 presents the experiments with the ERBFN for the Yen/Dollar time series.

Table 2 presents the experiments of T1FS Integration of the Yen/Dollar time series.

Table 3 shows the IT2FS experiments, with different degrees of uncertainty for the Yen/Dollar time series.

Figure 6 presents the plot of the best results ERBFN, that is the actual data against the predicted data for the Yen/Dollar.

Table 4 presents the experiments of ERBFN for the Yen/Mexican pesos time series.

Table 5 presents the experiments of the T1FS for Yen/Mexican pesos time series.

Table 6 shows the IT2FS experiments, with different degrees of uncertainty for the Yen/Mexican pesos time series.

Figure 7 presents the plot of the best results ERBFN, that is the actual data against the predicted data for the Yen/Mexican pesos.

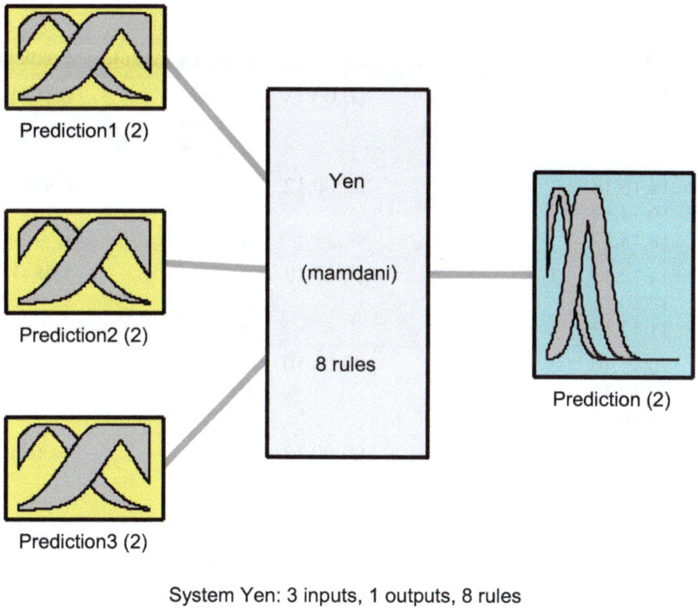

Fig. 4 IT2FS structure

1. If (Pd1 is Pe1L) and (Pd2 is Pe2L) and (Pd3 is Pe3L) then (Pd is Lw)
2. If (Pd1 is Pe1H) and (Pd2 is Pe2H) and (Pd3 is Pe3H) and then (Pd is Hgh)
3. If (Pd1 is Pe1L) and (Pd2 is Pe2L) and (Pd3 is Pe3H) then (Pd is Lw)
4. If (Pd1 is Pe1H) and (Pd2 is Pe2H) and (Pd3 is Pe3H) and then (Pd is Hgh)
5. If (Pd1 is Pe1L) and (Pd2 is Pe2H) and (Pd3 is Pe3H) then (Pd is Hgh)
6. If (Pd1 is Pe1H) and (Pd2 is PeL) and (Pd3 is Pe3H) then (Pd is Lw)
7. If (Pd1 is Pe1L) and (Pd2 is Pe2H) and (Pd3 is Pe3L) then (Pd is Lw)
8. If (Pd1 is Pe1H) and (Pd2 is Pe2L) and (Pd3 is Pe3H) then (Pd is Hgh)

Fig. 5 IT2FS rules

4 Conclusions

Radial Function Neural Networks are an excellent architecture of time series prediction, and we conclude that this network provides us with good results, as well as the integration with type-1 fuzzy logic produced good results, as well as the integration with T1FS and IT2FS, varying the parameters until good results were obtained. As future work, we propose the optimization of the architecture of the ensemble radial basis network, in terms of the number of modules, layers, and number of neurons

Table 1 ERBFN results

No	NL	NN	Number of delays	Duration	Error by module	Prediction Error
1	3	14,16	3	00:00:13	0.06973 0.06808 0.06998	0.1614
2	3	14,16,18 16,14,19 18,15,17	3	00:00:12	0.0423 0.6998 0.6997	0.9753
3	3	14,12,13 12,10,15 11,13,12	3	00:00:12	0.5510 0.6998 0.6998	1.4841
4	3	10,10,10 12,10,12 11,10,12	3	00:00:10	0.0421 0.6853 0.6998	0.9607
5	3		3	00:00:12	0.1109 0.7226 0.6998	1.0688
6	3	8,9,10 9,12,11 8.9.11	3	00:00:11	0.7000 0.6998 0.2426	1.4876
7	3	10,10,10 10,12,10 10,9,10	3	00:00:11	0.0422 0.6991 0.6998	0.9746
8	3	11,10,9 10,11,10 10,11,10	3	00:00:09	0.6864 0.0419 0.7070	0.9649
9	3	11,11,12 13,12,12 10,1312	3	00:00:11	0.7137 0.6998 0.04919	1.42575
10	3	11,13,12 13,11,12 10,12,12	3	00:00:10	0.6340 0.6902 0.6998	1.5575

Table 2 T1FS for the Yen/Dollar data

No	T1FS
1	0.3815
2	0.3905
3	0.3907
4	0.3905
5	0.3908
6	0.3863
7	0.3722
8	0.3888
9	0.3887
10	0.3941

Table 3 IT2FS for Yen/ Dollar data

Experiments	0.3	0.4	0.5
1	0.2320	0.2321	0.2223
2	0.2318	0.2322	0.2328
3	0.2315	0.2332	0.2339
4	0.2314	0.233	0.2338
5	0.2328	0.2330	0.2338
6	0.2333	0.2225	0.2231
7	0.2324	0.2325	0.2336
8	0.2362	0.2372	0.2376
9	0.2358	0.2362	0.2371
10	0.2378	0.2382	0.2387

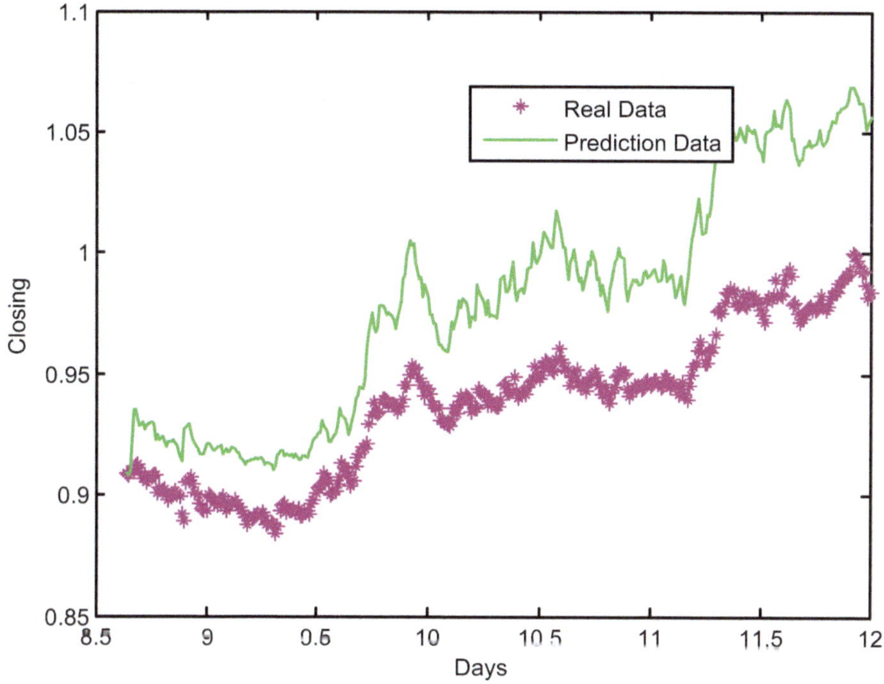

Fig. 6 Graph of the Yen/ Dollar

per layer, which could be done with metaheuristics, as in [32–35]. Another future work could be to use type-3 fuzzy logic in the integration, as in [36–39]. Finally, the approach could be utilized in other kinds of applications [40–44].

Table 4 Yen/Mexican pesos time series results of the EBFNN

No	NL	NN	Number of delays	Duration	Error by module	Prediction error
1	3	18,15,16 18,15,17 16,16,18	3	00:00:13	0.0099 0.0100 0.0099	0.0232
2	3	18,15,18 18,15,15 16,16,15	3	00:00:06	0.0099 0.0099 0.0088	0.0228
3	3	13,15,15 14,15,14 16,15,15	3	00:00:06	0.0099 0.0098 0.0104	0.0231
4	3	13,12,15 14,13,14 12,15,13	3	00:00:06	0.0099 0.0098 0.0098	0.0230
5	3	10,10,12 12,13,10 10,14,13	3	00:00:06	0.0097 0.0102 0.0097	0.0235
6	3	12,12,10 11,11,12 12,10,11	3	00:00:07	0.0097 0.0098 0.0098	0.0227
7	3	11,12,10 12,11,10 12,10,10	3	00:00:06	0.0098 0.0087 0.0092	0.0217
8	3	10,12,10 11,11,12 10,10,11	3	00:00:04	0.0098 0.0098 0.0099	0.0231
9	3	11,11,10 11,11,11 11,11,10	3	00:00:05	0.0099 0.0098 0.0668	0.0668
10	3	11,11,11 11,11,11 10,11,11	3	00:00:06	0.0099 0.0099 0.1394	0.0662

Table 5 T1FS in Yen/Mexican pesos

No	T1FS
1	0.2476
2	0.2472
3	0.2471
4	0.2471
5	0.2477
6	0.2297
7	0.2541
8	0.2248
9	0.2488
10	0.2397

Table 6 IT2FS for Yen/ Mexican pesos

Experiments	0.3	0.4	0.5
1	0.3320	0.3321	0.3223
2	0.3318	0.3322	0.3328
3	0.3315	0.3332	0.3339
4	0.3314	0.3320	0.3338
5	0.3328	0.3330	0.3338
6	0.3333	0.3225	0.3231
7	0.3324	0.3325	0.3336
8	0.3362	0.3372	0.3376
9	0.3358	0.3362	0.3371
10	0.3378	0.3382	0.3387

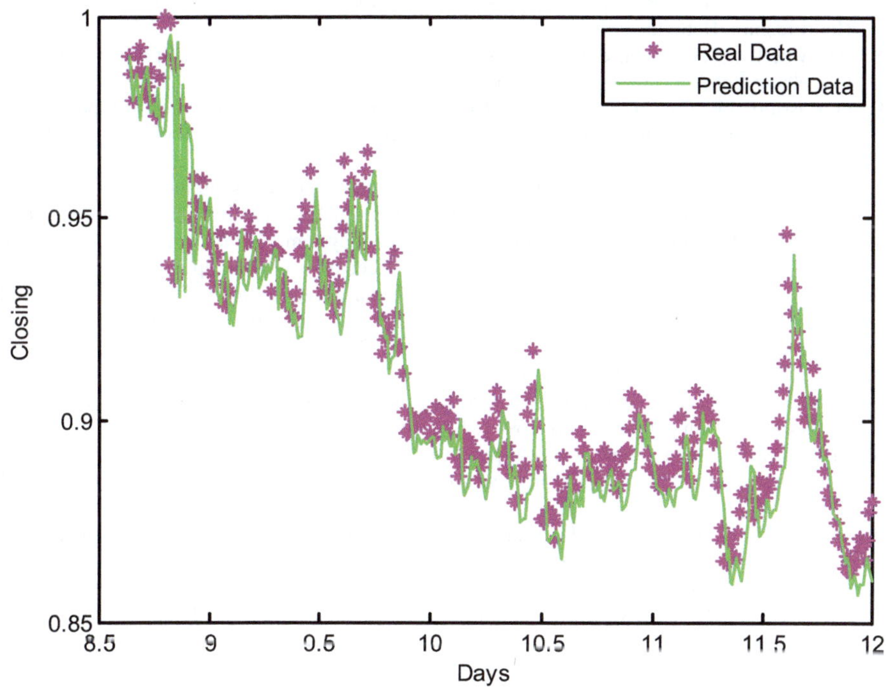

Fig. 7 Plot of the Yen/ Mexican Pesos

Acknowledgements We would like to express our gratitude to the CONACYT, Tijuana Institute of Technology for the facilities and resources granted for the development of this research.

References

1. A. Famelis Donas, G. Galanis, Comparative study of feedforward and radial basis function neural networks for solving an environmental boundary value problem. Appl. Mathem. **16**, 1–11 (2022)
2. Y. Bodyanskiy, A. Pirus, A. Deineko, Multilayer radial basis function net and its learning, IEEE 15th International Conference on Computer Sciences and Information Technologies, pp. 92–95 (2020)
3. G. Preethi Sathiyapriya, S. Anita Shanth, Radial basis function bipolar fuzzy neural network. Mater. Today: Proc., 1–6 (2022)
4. Z. Sanpeng, F. Renzhong, A variable projection method for the general radial basis function neural network. Appl. Mathem. Comput. 1–17 (2023)
5. W. Ji, F. Leichao, D. Guangzhong, L. Mingqiang, "State of health estimation of lithium-ion battery with improved radial basis function neural network. Energy **262**, 1–17 (2023)
6. P.D. Brockwell, R.A. Davis, *Introduction to time series and forecasting* (Springer-Verlag, New York, 2002), pp.1–219
7. N. Davey, S. Hunt, R. Frank, *Time series prediction and neural networks* (University of Hertfordshire, Hatfield, 1999)
8. P. Cowpertwait, A. Metcalfe, Time series. Introductory Time Series with R., Springer Dordrecht Heidelberg London New York, pp. 2–5 (2009)
9. O. Castillo, J.R. Castro, P. Melin, A. Rodriguez-Diaz, Application of interval type-2 fuzzy neural networks in non-linear identification and time series prediction. Soft. Comput. **18**(6), 1213–1224 (2014)
10. O. Castillo, P. Melin, Hybrid intelligent systems for time series prediction using neural networks, fuzzy logic, and fractal theory. Neural Netw. IEEE Trans. **13**(6), 1395–1408 (2002)
11. O. Castillo, P. Melin, Comparison of Hybrid Intelligent Systems, Neural Networks and Interval Type-2 Fuzzy Logic for Time Series Prediction. Proceedings IJCNN, pp. 3086–3091 (2007)
12. J. Castro, O. Castillo, P. Melin, O. Mendoza, A. Rodríguez, An interval type 2 fuzzy neural network for chaotic time series prediction with cross- validation and Akaike test", Soft Computing for Intelligent Control and Robotics, pp. 269–285 (2011)
13. M. Pulido, P. Melin, O. Mendoza, Optimization of ensemble neural networks with type-1 and interval type-2 fuzzy integration for forecasting the taiwan stock exchange. Advances in Data Analysis with Computational, pp. 169–181 (2018)
14. M. Ramirez, P. Melin, A new interval type-2 fuzzy aggregation approach for combining multiple neural networks in clustering and prediction of time series. Int. J. Fuzzy Syst. **25**(3), 1077–1104 (2023)
15. O. Castillo, J. Castro, P. Melin, Interval type-3 fuzzy aggregation of neural networks for multiple time series prediction: the case of financial forecasting. Axioms **11**(6), 251 (2022)
16. M.A. Koops, L.A. Giraldeau, Producer-scrounger foraging games instarlings: A test of rate maximizing and risk-sensitive model Anmal Behavior, pp. 773–783 (1996). https://doi.org/10.1006/anbe.1996.0082
17. J. Krause, G.D. Ruxton, Living in groups. Oxford: Oxford University Press (2002)
18. T.R. Anderson, *Biology of the ubiquitous house sparrow: from genes populations* (Oxford University Press, Oxford, 2002)
19. S.L. Lima, L.M. Dill, Behavioral decisions made under the risk of predation: a review and prospectus. Can. J. Zool. **2002**, 619–640 (1990)
20. H.R. Pulliam, On the advantages of flocking. J. Theoret. Call Biol. **38**, 419–422 (1973). https://doi.org/10.1016/0022-5193(73)90184-7
21. H.R. Pulliam, G.H. Pyke, T. Caraco, H.R. Pulliam, The scanning behavior of juncos: a game theoretical approach. J. Theor. Biol. 89- 103 (1982). doi:https://doi.org/10.1016/0022-5193(82)90289–2
22. G. Beauchamp, Group-size effects on vigilance: a search for mechanisms. Behav Proc, 111–121 (2006). https://doi.org/10.1016/S0376-6357(03)00002-0

23. B.A. Bednekoff, S.L. Lima, Randomness, chaos and confusion in the study of antipredator vigilance. Trends in Ecology and Evolution, pp. 284–297 (2006)
24. M. Pulido, P. Melin, Ensemble recurrent neural networks for complex time series prediction with integration methods. Fuzzy Logic Hybrid Extensions of Neural and Optimization Algorithms, pp. 71–83 (2021)
25. A. Sherstinsky, Fundamentals of recurrent neural network (RNN) and long short-term memory (LSTM) network. Phys. D: Nonl. Phenom., 1–28 (2020)
26. J. Mendel, Uncertain rule-based Fuzzy logic systems. Introduction a new direction. Prentice-Hall, Inc., 213–231 (2001)
27. O. Castillo, P. Melin, Simulation and forecasting complex economic time series using neural networks and fuzzy logic. Proc. Intern. Neural Netw. Conf. **3**, 1805–1810 (2001)
28. O. Castillo, P. Melin, Type-2 Fuzzy Systems. Type-2 Fuzzy logic Theory and Application, Ed. Springer, pp. 30–43 (2008)
29. O. Castillo, *Type-2 fuzzy logic in intelligent control applications* (Springer, Berlin, Germany, 2012)
30. N. Karnik, J.M. Mendel, Introduction to type-2 fuzzy logic systems. IEEE Trans. Signal Proc. **2**, 915–920 (1998)
31. Yen Database: https://mx.investing.com/commodities/crude-oil-historical- data (2023)
32. F. Valdez, J.C. Vazquez, P. Melin, O. Castillo, Comparative study of the use of fuzzy logic in improving particle swarm optimization variants for mathematical functions using co-evolution. Appl. Soft Comput. **52**, 1070–1083 (2017)
33. D. Sanchez, P. Melin, O. Castillo, A grey wolf optimizer for modular granular neural networks for human recognition. Comput. Intell. Neurosci. **2017**(2017). https://doi.org/10.1155/2017/4180510
34. O. Castillo, E. Lizzarraga, J. Soria, P. Melin, F. Valdez, New approach using ant colony optimization with ant set partition for fuzzy control design applied to the ball and beam system. Inf. Sci. **294**, 203–215 (2015)
35. L. Amador-Angulo, O. Mendoza, J.R. Castro, A. Rodriguez-Diaz, P. Melin, O. Castillo, Fuzzy sets in dynamic adaptation of parameters of a bee colony optimization for controlling the trajectory of an autonomous mobile robot. Sensors **16**(9), 1458 (2016)
36. O. Castillo, J.R. Castro, P. Melin, *Interval type-3 fuzzy systems: theory and design* (Springer, Cham, Switzerland, 2022)
37. J.T. Rickard, J. Aisbett, G. Gibbon, Fuzzy subsethood for fuzzy sets of type-2 and generalized type-n. IEEE Trans. Fuzzy Syst. **17**(1), 50–60 (2009)
38. O. Castillo, Towards finding the optimal *n* in designing type-*n* fuzzy systems for particular classes of problems: a review. Appl. Comput. Math. **17**(1), 3–9 (2018)
39. O. Castillo, J.R. Castro, P. Melin, Interval type-3 fuzzy control for automated tuning of image quality in televisions. Axioms **11**, 276 (2022)
40. B. González, F. Valdez, P. Melin, G. Prado-Arechiga, Fuzzy logic in the gravitational search algorithm for the optimization of modular neural networks in pattern recognition. Expert Syst. Appl. **42**(14), 5839–5847 (2015)
41. K. Tai, A.-R. El-Sayed, M. Biglarbegian, C.I. Gonzalez, O. Castillo, S. Mahmud, Review of recent type-2 fuzzy controller applications. Algorithms **9**(2), 39 (2016)
42. P. Melin, O. Castillo, Intelligent control of a stepping motor drive using an adaptive neuro-fuzzy inference system. Inf. Sci. **170**(2–4), 133–151 (2005)
43. P. Melin, O. Castillo, A new method for adaptive control of non-linear plants using type-2 fuzzy logic and neural networks. Int. J. Gen. Syst. **33**(2–3), 289–304 (2004)
44. P. Melin, O. Castillo, Adaptive intelligent control of aircraft systems with a hybrid approach combining neural networks, fuzzy logic and fractal theory. Appl. Soft Comput. **3**(4), 353–362 (2003)

Evolutionary Computing

Hypervolume Indicator as an Estimator for Adaptive Operator Selection in an On-Line Multi-objective Hyper-heuristic

Jorge A. Soria-Alcaraz, Gabriela Ochoa, Marco A. Sotelo-Figueroa, and Andres Espinal

1 Introduction

A multi-objective optimization problem involves optimizing multiple conflicting objectives. Unlike single-objective optimization problems, where the goal is typically to find a single solution that minimizes or maximizes a single objective, multi-objective optimization aims to find a set of Pareto-optimal solutions. i.e. $f(x) = (f_1(x), \ldots, f_M(x))$ objective functions where $M \geq 2$ is the number of objectives. This type of problem can be frequently found in real-world settings.

Hyper-heuristics can be conceived as high-level algorithms that select or generate low-level heuristics to solve optimization problems [1, 2]. Hyper-heuristics try to reduce the required expertise to construct/select optimization methods [3, 4].

Hyper-heuristics have been applied to multi-objective domains with encouraging results. Recent works include applications in timetabling [5], the travelling salesman problem [6], water distribution network design [7], software module clustering [8], Shelf-space allocation [9] and wind farm layout optimization [10]. However, there are very few works related to online selection hyper-heuristics over multi-objective domains.

The MOEA/D (*multi-objective evolutionary algorithm based on decomposition*) [11] decomposes a multi-objective optimization problem into a number of scalar optimization sub-problems that are optimized simultaneously. In this work, we use MOEA/D as the high-level strategy for our online selection hyper-heuristic.

J. A. Soria-Alcaraz (✉) · M. A. Sotelo-Figueroa · A. Espinal
Departamento de estudios organizacionales, Universidad de Guanajuato, Guanajuato, Mexico
e-mail: jorge.soria@ugto.mx

M. A. Sotelo-Figueroa
e-mail: masotelof@ugto.mx

A. Espinal
e-mail: aespinal@ugto.mx

G. Ochoa
Computing and Mathematics Department, University of Stirling, Stirling, UK
e-mail: gabriela.ochoa@cs.stir.ac.uk

P. Melin and O. Castillo (eds.), *New Directions on Hybrid Intelligent Systems Based on Neural Networks, Fuzzy Logic, and Optimization Algorithms*, Studies in Computational Intelligence 1146, https://doi.org/10.1007/978-3-031-53713-4_9

During an online selection, a particular operator is picked from a pool to proceed with the search process. The goal is to utilize the most efficient heuristic at a specific stage of execution. This leads to an automated approach that consecutively chooses heuristics to enhance the existing solution.

Online selection hyper-heuristics have the goal of applying the most efficient or effective heuristic at a certain execution point to guide a search process. Selecting the best potential heuristic for continuing a search is not an easy task because the quality of each heuristic changes dynamically during the search process. That is why it is necessary to record the occasioned improvement (or the lack of) of the last applied heuristic. Then at later search stages, this record will be helpful when deciding the following heuristic to be applied.

In this work, the high-level algorithm selects an adequate heuristic using an Adaptive Operator Selection (AOS) approach. AOS combines two cooperating strategies: (i) *operator selection* which selects the next operator to be applied based on its estimated potential success; and (ii) *credit assignment* which assigns a reward value to a specific heuristic based on the observed performance on the last iterations.

Usually, this reward value is calculated as the fitness improvement brought by a given operator or low-level heuristic [12]. Our proposal is to use the Hypervolume indicator difference between an offspring population (affected by an operator) and a parent one as a fitness improvement metric. This idea allows us to apply single-objective AOS techniques to evolutionary multi-objective domains.

The Hypervolume indicator (Hv) [13] is an interesting metric of the Pareto dominance relation. An optimal $\mu-$distribution of this indicator contains only Pareto-optimal solutions. This means that a solution set that maximizes the Hypervolume (Hv) indicator is guaranteed to cover the optimal Pareto front [14].

To test and compare our algorithm, we apply it to well-known optimization benchmarks; namely, ZDT and DTLZ benchmarks. The next section describes important concepts like the high-level MOEA/D algorithm, Hypervolume indicator theory and AOS theory. Our proposal is detailed in Sect. 3. Section 4 summarizes the empirical setup as well as the solution metrics, and Sect. 5 reports and analyzes our results. Finally, Sect. 6 summarizes our conclusions and discusses future work.

2 Relevant Concepts

2.1 On-Line Hyper-heuristic

Hyper-heuristics are conceptualized as *heuristics to choose heuristics* [4]. These methodologies reduce the need for a human expert raising the level of generality at which search methodologies can operate. A recent definition considers a hyper-heuristic as a "*automated methodologies for selecting or generating heuristics to solve computational search problems*" [2].

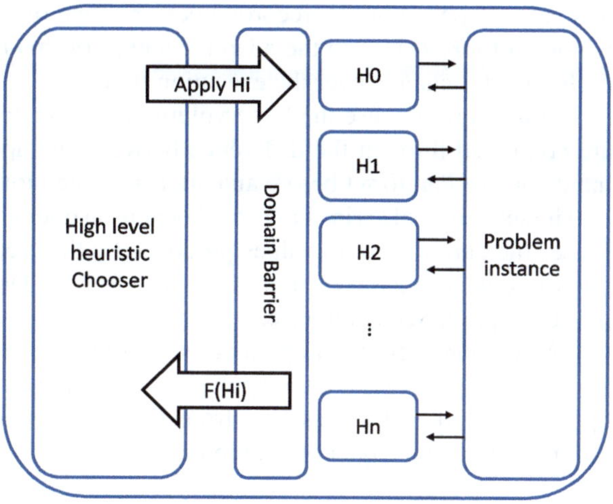

Fig. 1 General framework of a selection hyper-heuristic based on Cowling et al. [4]

Figure 1 illustrates the general selection hyper-heuristic framework; the *Domain barrier* is an important element that separates a general search strategy (high level) from the actual problem (low level). This separation layer allows, in theory, that the same high-level search strategy can be applied to a range of different domains. In addition, a pool of problem-dependent heuristics is needed from which, at each iteration, a single heuristic will be selected by the high-level strategy to be applied to the incumbent solution.

Online hyper-heuristics can be considered learning algorithms since they use feedback from the current state of search to rank heuristics in terms of their observed performance. State of art distinguishes between *online* and *offline* learning hyper-heuristics [15]. Online learning takes place when solving a problem instance, while offline learning requires a training process before execution of the hyper-heuristic. In this work, we use online learning through *adaptive operator selection.*

2.2 *Adaptive Operator Selection*

Adaptive operator selection (AOS) is a method for dynamically choosing, at each iteration, the operator that is most likely to result in an improvement in the solution quality. This concept has been used within metaheuristics and evolutionary algorithms [16, 17], and has also produced positive outcomes within hyper-heuristics [5]. In this study, we employ AOS within MOEA/D-DRA to improve performance during the mutation phase.

Two mechanisms are used by AOS to select a low-level operator: *credit assignment* and *selection rule*. Credit assignment gives rewards to operators based on their recent performance. A high reward for an operator indicates that during its recent application, the operator triggered positive changes in the solution quality. Usually, this value depends on the recorded performance of a given operator during a time window of size l. Due to the adaptive nature of this mechanism, reward values for each operator change over the execution time.

In this work, we use the hypervolume indicator for credit assignment. Rewards are computed through the difference between the application of the hypervolume indicator to a Pareto set before and after the execution of a mutation operator.

The selection rule is in charge of choosing an operator to apply at further iterations. It uses the current reward values produced by the credit assignment mechanism. In this work, we use the Dynamic Multi-Arm Bandit (DMAB) previously documented in [16, 17] as a Selection Rule.

DMAB considers each operator as an arm. Let $n_{i,t}$ denote the number of times the ith arm/operator has been applied, and $\hat{p}_{i,t}$ the empirical reward it has received up to time t. At each time step t, from K alternative arms, the algorithm selects the arm maximising the quantity computed by Eq. 1.

$$\hat{p}_{j,t} + C\sqrt{\frac{2log(\sum_{i=1}^{K} n_{i,t})}{n_{j,t}}} \tag{1}$$

A scaling factor C is necessary in order to properly balance the trade-off between selecting the operator with the best empirical behaviour (Eq. 1) and giving other operators the opportunity to be selected (second term in Eq. 1). We used the value suggested in [18] ($\delta = 0.15$) for all the experiments. We selected DMAB as the selection rule due to its hybridization with the Page-Hinkle test, and its good performance achieved in previous works [5, 17, 18].

2.3 *MOEA/D-DRA*

A multi-objective problem can be broken down into numerous single-objective problems using the Multiobjective Evolutionary Algorithms based on Decomposition (MOEA/D) technique [11]. Following that, each subproblem's aim is taken into account along with all the other objectives of the original multi-objective problem in a linear or nonlinear aggregation function. The simultaneous optimization of each subproblem relies mostly on the information of its nearby subproblems. These subproblems may require different amounts of computational resources so it is reasonable to apply a strategy to share efficiently the available computational resources to each part of the problem [19].

To address this phenomenon, we use a variant of the classic MOEA/D algorithm; Namely, MOEA/D-DRA (MOEA/D with Dynamical Resource Allocation) as a high-level strategy. In the MOEA/D-DRA algorithm, a utility value for each subproblem is defined and computed, and computational efforts are distributed to these subproblems [19].

The MOEA/D-DRA algorithm uses crossover and mutation operators to guide the search. Several crossover operators have been proposed in the literature with encouraging results. In this paper, we use the Simplex Crossover operator as proposed in [20].

The impact of mutation operators, on the other hand, has not been studied exhaustively in the MOEA/D-DRA algorithm. In this work, we focus on the adaptive selection of mutation operators through the application of a Dynamic Multi-Arm Bandit strategy.

2.4 *Hypervolume Indicator*

The hypervolume indicator quantifies the volume of the space that is dominated by a set of solutions. Essentially, it measures the size of the space for which the given set of solutions is better than or equal to any other potential solutions.

- **Pareto Front**: Given a set S of non-dominated solutions in a d-dimensional objective space, the Pareto front is defined as the set of all objective vectors of the solutions in S.
- **Reference Point**: To compute the hypervolume, a reference point r is needed, which is usually chosen to be worse than any of the solutions in S for all objectives. This ensures that the hypervolume computation covers the entire region of interest.
- **Volume Calculation**: The hypervolume $HV(S)$ of a set of solutions S relative to a reference point r is defined as the volume of the union of all hyper-rectangles (or boxes) defined by each solution in S and the reference point r.
  Mathematically, the hypervolume $HV(S)$ of a set S with respect to a reference point r is given by:

  $$HV(S) = \text{Leb}\left(\bigcup_{s \in S} [s, r]\right)$$

  where $[s, r]$ represents the hyper-rectangle formed by solution s and the reference point r, and Leb denotes the Lebesgue measure, which, in simpler terms, computes the volume of the region [21].

The hypervolume has a few properties that make it particularly useful:

- **Pareto-dominance**: An increase in hypervolume indicates a set of solutions that dominates another set.
- **Cardinality**: If a solution set S' is obtained by adding non-dominated solutions to a set S, then $HV(S') > HV(S)$.
- **Uniformity**: It is sensitive to the distribution of solutions across the Pareto front. This is desirable as we often want a diverse set of solutions that are spread out across the Pareto front.

In essence, the hypervolume indicator transforms a multi-objective problem into a single-objective one consisting of finding a set of points maximizing the respective indicator I.

3 Methodology

In this section, we detail our approach using the hypervolume indicator (Hv) as an estimator for adaptive operator selection in an online multiobjective hyper-heuristic. As seen in Sect. 2.2, an adaptive operator selection scheme consists of two components: a *credit assignment* mechanism and a *selection rule*. It is within the credit assignment mechanism that the Hypervolume indicator is used as an estimator for the quality of a given operator. This approach is detailed in Sect. 3.2. Section 3.1 describes the high-level MOEA/D-DRA algorithm used.

3.1 High-Level MOEA/D-DRA Strategy

MOEA/D algorithms minimize N objective functions built from a single multi-objective problem. The variant that we utilize corresponds to the MOEA/D-DRA [cec09], where a utility π^i is defined for each subproblem created for the MOEA/D framework, The idea is to assign computational resources accordingly to the observed difficulty of each subproblem $i = 1, \ldots, N$.

Algorithm 1 details our implementation of MOEA/D-DRA using the JMETAL framework source code [22]. Our AOS proposal is applied in line 11 as an adaptive mutation operator. This line executes our Algorithm 2 to dynamically select the mutation operator, and thus enhance the search process.

3.2 Hypervolume Indicator as an Operator Quality Metric

Hypervolume indicator theory seen in Sect. 2.4 allows us to handle a multi-objective problem by means of the maximization of a single-objective that is the Hv value. Maximizing this value theoretically corresponds to finding the optimal Pareto front. As described in Sect. 2.2, a *credit assignment* mechanism usually utilizes a single-objective fitness function evaluation before and after the application of a given operator in order to identify how good it was in terms of performance. The key idea in our work is to apply a Hv indicator evaluation to a solution set before and after the execution of a given operator and, therefore, assign a reward according to the maximization degree in the Hv value. This allows us to adaptively select operators in multi-objective domains as long as we have a solution set (guaranteed by the use of a population-based algorithm like the MOEA/D-DRA) and the hypervolume indicator.

Algorithm 1 High-Level MOEA/D-DRA Algorithm.

Require: F : fitness function, $sample$: number of samples, size of parent pool (1), λ : size of the offspring pool, op : Heuristic or operator to sample.
$S \leftarrow$Generate InitialPopulation $S = (s_1, \ldots, s_n)$
$N \leftarrow$ Number of Subproblems.
$T \leftarrow$ Number of Weight Vectors
$EP \leftarrow 0$
Setting of Neighborhood and then work out the T closest weight vectors to each weight vector(λ_i)
Calculate the fitness values of the population.
Initialize $z = (z_1, \ldots, z_m)^r$
while Stopping criterion is not met **do**
    $S* \leftarrow Crossover(S)$
    $s \leftarrow SelectSolutiontoMutate(S)$
    $s' \leftarrow Mutation(hv, S, s)$
    $UpdatePopulation(S, S*, s'])$
    $Update(z)$
    Update Neighboring Solution
    Update EP
end while
return $S = 0$

Algorithm 2 Mutation phase with AOS and Hv indicator for multi-objective domains.

Require: Hv : Hypervolume indicator function, S : solution set, $s \in S$ solution to mutate, Op list of operators.
1: $Hv_0 \leftarrow Hv(S)$
2: $S^* \leftarrow S - s$
3: $op \leftarrow SelectHeuristic(Op)$ (selection rule)
4: $s' \leftarrow apply(op, s)$
5: $S^* \leftarrow S^* + s'$
6: $Hv_1 \leftarrow Hv(S^*)$
7: $UpdateRewards(op, Hv_1 - Hv_0)$ (credit assignment)
8: **return** $s' = 0$

Algorithm 2 details the mutation phase of our high-level MOEA/D-DRA strategy. Using the HV evaluation at lines 1 and 6 of Algorithm 2 we can assign a reward to the last operator used according to the maximization of this indicator. This allows our credit assessment strategy (DMAB) and selection rule (extreme) to work without further changes as seen in Sect. 2.2.

4 Experiments

Our experimental setup is detailed in this section along with the parameters, algorithms, and statistical tests used.

4.1 Test Problems

As test instances, we use the multi-objective [ZDT] problems [23]. This benchmark consists of 6 minimizing functions with $d = 2$ dimensions and the following characteristics.

ZDT1: Convex Pareto optimal front.
ZDT2: Non-convex counterpart to ZDT1.
ZDT3: Pareto optimal front consists of several non-contiguous convex parts.
ZDT4: Contains 21^9 local Pareto optimal fronts. (designed to test algorithms ability to deal with multi-modality).
ZDT6: Pareto-optimal solutions are non-uniformly distributed along the global Pareto front.

DTLZ problems proposed by [24] are also used as test instances, in this second benchmark, test instances are used with $d = 3$. Characteristics of these instances are as follows.

DTLZ1: Linear Pareto-optimal front.
DTLZ2: Generic Sphere Pareto-optimal front.

4.2 Algorithms and Parameter Settings

In order to use our MOEA/D-DRA with AOS mutation phase based on the hypervolume indicator, we need a pool of low-level mutation heuristics; the selected pool is the following:

Simple Random Mutation: Mutates a single variable of a given solution.
Uniform Mutation: Mutates a variable following a uniform at random probably.
Polinomial Mutation: The classic NSGA-II polynomial mutation with a static distribution index.
Non-Uniform Mutation: Mutates a variable following a normal distribution.

We executed 50 independent experiments with a population of 300 solutions. The number of evaluations of each objective function was limited to 25,000 across our benchmark. We use an i7core CPU along with 16GB RAM using the Java language. Finally, our configurations were:

No Selection (NS): Algorithm variant with a fixed polynomial mutation
Random Selection (RS): Algorithm variant with uniform at random heuristic selection.
Hypervolume AOS (HV AOS): Algorithm variant with HV indicator for credit assignment.

Table 1 Mean results on selected ZDT instances

Algorithm	ZDT 1	ZDT 2	ZDT 3	ZDT 4
NS	EpI: 1.04e-2 Hv: 0.65 GD: 1.6e-3	EpI: 1.2e-3 Hv: 0.32 GD: 1e-3	EpI: 8.9e-3 Hv: 1.04 GD: 1.3e-3	EpI: 0.014 Hv: 0.65 GD: 3e-3
RS	EpI: 6.9e-3 Hv: 0.66 GD: 3.5e-3	EpI: 1.1e-2 Hv: 3.2e-1 GD: 7.2e-4	EpI: 7.8e-3 Hv: 1.04 GD: 1.3e-3	EpI: 0.35 Hv: 0.53 GD: 2.1e-2
HV AOS	EpI: **0.58** Hv: **4e-3** GD: **0.63**	EpI: **1.69** Hv: **0** GD: **0.79**	EpI: **0.62** Hv: **0.42** GD: **0.34**	EpI: **6.71** Hv: **0** GD: **13.26**

Table 2 Min and mean results on ZDT and DTLZ instances using IGD indicator

IGD	No selection		Random selection		Hv AOS	
Problem	Median	Best	Median	Best	Median	Best
ZDT1	0.0116	0.0105	0.0121	0.0165	0.0106	**0.0082**
ZDT2	0.0225	0.0177	0.0105	0.0080	0.0136	**0.0100**
ZDT3	0.0272	0.0176	0.0148	0.0121	0.0130	**0.0170**
ZDT4	0.2474	0.0747	0.0133	0.0176	0.0197	**0.0138**
ZDT6	0.0149	0.0122	0.0069	0.0062	0.0044	**0.0092**
DTLZ1	0.3085	0.0737	0.256	0.0492	0.269	**0.0558**
DTLZ2	0.0319	0.0243	0.0351	0.0307	0.2901	**0.0287**

We use 3 metrics to compare the quality of the last population obtained by the MOEA/D-DRA variants. Namely, Epsilon Indicator (EpI), hypervolume (Hv), and Generational Distance (GD) and Inverse Generational Distance(IGD) [25].

5 Results and Discussion

In this section, we detail the results obtained by our algorithm configurations.

Table 1 shows the mean of 35 executions of our algorithms with the parameter configuration reported in Sect. 4.2. This experiment shows the better performance (bold font) of our approach which uses a Hypervolume indicator as credit assignment in adaptive operator selection. This experiment is important since we have evidence of two key ideas: (A) The improvement in performance because of the usage of adaptive operator selection against a uniform at random selection of operators and the no selection of operators at all. (B) The Hypervolume indicator can guide operator selection over ZDT instances.

Table 2 shows the median and best results achieved by our approach against other configurations. Consistently, the best results (lower values in bold font) are obtained by adaptive operator selection using Hypervolume as a credit assignment. The methodology proposed in this work also makes it possible to apply any single-objective operator selection technique to a multi-objective problem without further modifications.

6 Conclusions

This work has presented a methodology to use the Hypervolume indicator as a credit assignment rule for adaptive operator selection in an online multi-objective hyper-heuristic. The Hypervolume metric offers a simple yet effective way to apply single-objective hyper-heuristics to multi-objective problems.

An adaptive operator selection methodology has two key components: selection rule and credit assignment. In this work, we use dynamic multi-armed bandit as the selection rule and Hypervolume indicator as credit assignment. This work has shown that single objective selection rules can be used along with our proposal of a Hypervolume indicator for credit assignment.

Online hyper-heuristics were applied using the MOEA/D-DRA algorithm as a high-level algorithm. MOEA/D-DRA showed effective performance all across our benchmark dataset. Experiments gather encouraging evidence about the usage of adaptive operator selection in multi-objective instances. GD, IDG, Hv and Epsilon indicators have shown consistently that our approach performs better than no selection or random selection of low-level mutation operators.

For future work, alternative high-level multi-objective algorithms can be tested as well as low-level mutation operators. Experimentation with more complex instances or real-world problem instances is also a further natural step for this methodology. Finally, other selection rules like *adaptive pursuit* or *parallel matching* can be applied to this methodology.

Acknowledgements The authors wish to thank the University of Guanajuato and the University of Stirling.

References

1. P. Auer, H. Burgsteiner, W. Maass, A learning rule for very simple universal approximators consisting of a single layer of perceptrons. Neural Netw. **21**(5), 786–795 (2008)
2. E.K. Burke, M. Gendreau, M. Hyde, G. Kendall, G. Ochoa, E. Ozcan, R. Qu, Hyper-heuristics: a survey of the state of the art. J. Oper. Res. Soc. **64**(12), 1695–1724 (2013)
3. E.K. Burke, M. Hyde, G. Kendall, G. Ochoa, E. Özcan, J. Woodward, *Handbook of Metaheuristics*, vol. 146 of International Series in Operations Research & Management Science. Springer, Ch. A Classification of Hyper-heuristic Approaches, pp. 449–468, chapter 15 (2010)

4. P. Cowling, G. Kendall, E. Soubeiga, A hyperheuristic approach to scheduling a sales summit, in *Practice and Theory of Automated Timetabling III*, ed. by E. Burke, W. Erben. Lecture Notes in Computer Science, vol. 2079 (Springer, Berlin Heidelberg, 2001), pp.176–190
5. J.A. Soria-Alcaraz, G. Ochoa, J. Swan, M. Carpio, H. Puga, E.K. Burke, Effective learning hyper-heuristics for the course timetabling problem. Eur. J. Oper. Res. **238**(1), 77–86 (2014)
6. Y. Hidayatul, A. Djunaidy, A. Muklason, Solving multi-objective vehicle routing problem using hyper-heuristic method by considering balance of route distances, in *2019 International Conference On Information And Communications Technology (ICOIACT)* (2019), pp. 937–942
7. K. McClymont, E. Keedwell, D. Savić, M. Randall-Smith, A general multi-objective hyper-heuristic for water distribution network design with discolouration risk. J. Hydroinform. **15**, 700–716 (2013)
8. A. Kumari, K. Srinivas, Hyper-heuristic approach for multi-objective software module clustering. J. Syst. Softw. **117**, 384–401 (2016)
9. E. Burke, J. Silva, E. Soubeiga, Multi-objective hyper-heuristic approaches for space allocation and timetabling. *Metaheuristics: Progress As Real Problem Solvers*, pp. 129–158 (2005)
10. W. Li, E. Özcan, R. John, Multi-objective evolutionary algorithms and hyper-heuristics for wind farm layout optimisation. Renew. Energy. **105**, 473–482 (2017)
11. Q. Zhang, H. Li, MOEA/D: a multiobjective evolutionary algorithm based on decomposition. IEEE Trans. Evol. Comput. **11**, 712–731 (2007)
12. N. Veerapen, J. Maturana, F. Saubion, An exploration-exploitation compromise-based adaptive operator selection for local search, in *Genetic and Evolutionary Computation Conference, GECCO 2012. ACM, 2012*, pp. 1277–1284
13. E. Zitzler, L. Thiele, Multiobjective optimization using evolutionary algorithms-a comparative case study, in *International Conference On Parallel Problem Solving From Nature* (1998), pp. 292–301
14. A. Guerreiro, C. Fonseca, L. Paquete, The hypervolume indicator: computational problems and algorithms. ACM Comput. Surv. (CSUR) **54**, 1–42 (2021)
15. E.K. Burke, M. Hyde, G. Kendall, G. Ochoa, E. Özcan, J.R. Woodward, *A Classification of Hyper-heuristic Approaches* (Springer, US, Boston, MA, 2010), pp.449–468
16. L.D. Costa, A. Fialho, M. Schoenauer, M. Sebag, Adaptative operator selection with dynamic multi-armed bandits, in *Proceedings of the Genetic and Evolutionary Computation Conference (GECCO 2008). ACM, 2008*, pp. 913–920
17. A. Fialho, L.D. Costa, M. Sebag, Analyzing bandit-based adaptive operator selection mechanisms. Ann. Math. Artif. Intell. **60**(1–2), 25–64 (2010)
18. A. Fialho, L. Costa, M. Schoenauer, M. Sebag, Dynamic multi-armed bandits and extreme value-based rewards for adaptive operator selection in evolutionary algorithms, in *Learning and Intelligent Optimization*, vol. 5851 of Lecture Notes in Computer Science (Springer Berlin Heidelberg, 2009), pp. 176–190
19. Q. Zhang, W Liu, H. Li, The performance of a new version of MOEA/D on CEC09 unconstrained MOP test instances, in *2009 IEEE Congress On Evolutionary Computation* (2009), pp. 203–208
20. W. Khan, Q. Zhang, MOEA/D-DRA with two crossover operators, in *2010 UK Workshop On Computational Intelligence (UKCI)* (2010), pp. 1–6
21. A. Auger, J. Bader, D. Brockhoff, E. Zitzler, Hypervolume-based multiobjective optimization: Theoretical foundations and practical implications. Theoret. Comput. Sci. **425**, 75–103 (2012)
22. J. Durillo, A. Nebro, jMetal: a Java framework for multi-objective optimization. Adv. Eng. Softw. **42**, 760–771 (2011). https://www.sciencedirect.com/science/article/pii/S0965997811001219
23. E. Zitzler, K. Deb, L. Thiele, Comparison of multiobjective evolutionary algorithms: empirical results. Evol. Comput. **8**, 173–195 (2000)
24. K. Deb, L. Thiele, M. Laumanns, E. Zitzler, Scalable test problems for evolutionary multiobjective optimization, in *Evolutionary Multiobjective Optimization: Theoretical Advances And Applications* (2005), pp. 105–145

25. C. Coello Coello, M. Reyes Sierra, A study of the parallelization of a coevolutionary multi-objective evolutionary algorithm, in *MICAI 2004: Advances In Artificial Intelligence: Third Mexican International Conference On Artificial Intelligence, Mexico City, Mexico, April 26-30, 2004. Proceedings 3, 2004*, pp. 688-697

Developing a Quantum Genetic Algorithm in MATLAB Using a Quantum Device on AWS

Sandra S. Rosales-Alvarado, Oscar Montiel, Ulises Orozco-Rosas, and Juan J. Tapia

1 Introduction

The "optimization problem" is a recurring challenge in various fields, including economics, engineering, and medicine, owing to its significant implications. The term "optimization problem" refers to the quest for the best solution among possible solutions to a problem. One approach to tackling this issue is through the application of genetic algorithms. Simultaneously, there has been noteworthy progress in quantum computing, with efforts to enhance the performance of established algorithms. Notably, researchers have started to employ quantum metaheuristics, specifically emphasizing their foundational components, such as genetic algorithms. This exploration holds the potential to yield improved outcomes in addressing optimization challenges across diverse domains.

Genetic algorithms (GAs) have been widely employed for optimization problems, as demonstrated by Mitchell [1]. The fundamental concept behind optimization is to derive superior solutions that address a problem or enhance a process. One approach to achieving this objective involves exploring variations of an initial concept and leveraging this information to refine the original idea [2]; in other words, we can use a heuristic or metaheuristic method to modify a solution or evolve a population

Sandra S. Rosales-Alvarado, Oscar Montiel, Ulises Orozco-Rosas and Juan J. Tapia, these authors contributed equally to this work.

S. S. Rosales-Alvarado · O. Montiel (✉) · J. J. Tapia
Instituto Politécnico Nacional, CITEDI IPN, Av. Instituto Politécnico Nacional, No. 1310, Colonia Nueva Tijuana, 22435 Tijuana, B.C., Mexico
e-mail: oross@ipn.mx

S. S. Rosales-Alvarado
e-mail: srosales@citedi.mx

J. J. Tapia
e-mail: jtapiaa@ipn.mx

U. Orozco-Rosas
CETYS Universidad, Av. CETYS Universidad, No 4, Fracc. El Lago, 22210 Tijuana, B.C., Mexico
e-mail: ulises.orozco@cetys.mx

P. Melin and O. Castillo (eds.), *New Directions on Hybrid Intelligent Systems Based on Neural Networks, Fuzzy Logic, and Optimization Algorithms*, Studies in Computational Intelligence 1146, https://doi.org/10.1007/978-3-031-53713-4_10

until we find the optimal solution. Evolutionary metaheuristics are based on the idea above. Unfortunately, finding optimal solutions to huge, complex problems can be very computationally expensive, spanning from days to even years. Quantum metaheuristics are based on promising ideas of combining classical metaheuristics with quantum algorithms to develop quantum evolutionary algorithms aimed at reducing computational time.

At present, quantum computing faces problems arising from various sources, posing a significant challenge in their development; reducing the noise is one of the main challenges. Common noise causes include external electromagnetic interference, environmental interactions, calibration errors, finite qubit lifetimes due to decoherence, gate imperfections, and tunneling effects. Additionally, readout errors may occur during qubit measurements due to quantum sensitivity. Quantum error correction is an active research area aimed at mitigating these noise sources to enhance the reliability of quantum computers for practical applications.

For the above reasons, it is known that the current state of quantum computing is the NISQ (Noisy Intermediate-Scale Quantum) era, which is marked by the ongoing development of hardware, with quantum computers featuring tens or hundreds of qubits under construction; however, these machines remain susceptible to errors stemming from quantum interference and noise. Despite these limitations, practical applications in fields such as chemistry, optimization, medicine, and quantum cryptography are emerging, among others.

Sukhpal Gill et al. in [3] mentioned that the proliferation of quantum algorithms developed and tested on NISQ devices has been swift. They point out that, before the 1990s, only a few quantum algorithms were advertised, but today, many proposals exist. These include Variation Quantum Algorithms, as exemplified by their application to the Variational Quantum Eigensolver (VQE), as well as a concentration on Algebraic Quantum Algorithms applicable to discrete logarithms and matrix product verification. Furthermore, they highlight variational techniques, such as quantum approximate optimization, as the focal point of their research.

Currently, several well-established universal quantum computing platforms utilize the circuit model as the foundational framework for representing and executing quantum operations. In this model, quantum operations are depicted as quantum gates, and computations are structured as sequences of these gates, similar to classical digital circuits. Some examples of quantum computers employing this programming paradigm include the IBM Quantum Experience [4], Google's Sycamore [5], and Rigetti's quantum computing platform [6].

Genetic algorithms are optimization algorithms inspired by the process of natural selection and genetics, where candidate solutions evolve over generations to find the best solution to a problem. Quantum computing, on the other hand, leverages the principles of quantum mechanics to perform computations. Quantum bits, also known as qubits, can exist in multiple states simultaneously, enabling quantum computers to handle extensive data concurrently and potentially solve specific types of problems at a significantly faster rate compared to classical computers.

A quantum genetic algorithm (QGA) is an evolutionary algorithm that combines principles from quantum computing and genetic algorithms to solve optimization problems [7]. In a QGA, the traditional binary representation of genetic algorithms is replaced with a quantum representation, where solutions are encoded using quantum bits (qubits) instead of classical bits. This allows for exploring multiple potential solutions simultaneously, taking advantage of the parallelism inherent in quantum systems. QGA can be used to solve complex or impossible problems using classical algorithms.

This work introduces a methodology for bridging the gap between MATLAB, quantum computing, and Amazon Web Services (AWS). This methodology provides access to various quantum devices from different companies and technologies, including systems utilizing ion traps and superconductors, through AWS.

Our primary focus is facilitating the execution of Quantum Genetic Algorithms (QGAs) using MATLAB in conjunction with AWS. To demonstrate the process, we have initially implemented a basic QGA. However, our methodology extends beyond the execution of simple QGAs; it can support the development and execution of QGAs with multiple variables and objectives. Furthermore, this versatile methodology can be expanded to run various algorithms in diverse fields, including artificial intelligence, cryptography, chemistry, and more.

The main concepts and general knowledge employed in this book chapter are in Sect. 2. In Sect. 3, we delve into the implemented methodology, detailing algorithms and functional aspects. Section 4 presents the results of applying this methodology using a simulator and a quantum device. Finally, in Sect. 5, we provide the conclusions and future work.

2 Theoretical Framework

This section presents the quantum computing fundamentals that will enable an understanding of Quantum Genetic Algorithms (QGAs). We will also provide an overview of what AWS is and its available tools for running computational algorithms on simulators and quantum computers, focusing on those algorithms implemented in MATLAB.

2.1 Quantum Computing Fundamentals

The basic unit of information in quantum computing is the *qubit*, represented as $|\psi\rangle$, which encodes two basis states: $|0\rangle$ and $|1\rangle$ as a linear combination of both states. Note that for representing the values "0" and "1", we are using Dirac notation (bra-ket notation). Kets can also be represented as complex vectors. For example, for the value "0", the corresponding ket and vector are $|0\rangle = \begin{bmatrix} 1 & 0 \end{bmatrix}^T$, and for the value $|1\rangle$, it is $\begin{bmatrix} 0 & 1 \end{bmatrix}^T$.

A bra is also essentially a complex vector, typically denoted by $\langle\psi|$; it is the complex conjugate transpose of the ket vector $|\psi\rangle$. The bra lives in the dual space of the ket. Hence, the bra $\langle\psi| = |\psi\rangle^{\dagger}$, where the self-adjoint symbol † (dagger symbol) represents the complex conjugate transpose operation.

The linear combination of two states $|0\rangle$ and $|1\rangle$ mentioned above is known as the *superposition principle* of quantum states, which means that one qubit can exist in any superposition of the two basis states. Equation (1) illustrates this relationship,

$$|\psi\rangle = \alpha|0\rangle + \beta|1\rangle \text{ with } \alpha, \beta \in \mathbb{C} \tag{1}$$

in this context, α and β represent the probability amplitudes, and both of these values are complex. When we measure a qubit, the likelihood of obtaining the outcome $|0\rangle$ is determined by the square of the absolute value of α, while the probability of getting the outcome $|1\rangle$ is determined by the square of the absolute value of β. Always, the sum of both probabilities must fulfill the condition $|\alpha|^2 + |\beta|^2 = 1$.

In classical computing, a register is a set of bits; similarly, in quantum computing, a quantum register is a set of qubits. However, significant differences exist between a quantum qubit register and a classical bit register. Qubits can exist in a superposition of states, allowing them to represent both "0" and "1" simultaneously, thereby enabling quantum computers to process multiple possibilities concurrently.

$$|\Psi\rangle = |\psi_1\rangle \otimes \ldots \otimes |\psi_n\rangle \tag{2}$$

$$|\psi_{101}\rangle = |101\rangle = |1\rangle \otimes |0\rangle \otimes |1\rangle = \begin{bmatrix} 0 \\ 1 \end{bmatrix} \otimes \begin{bmatrix} 1 \\ 0 \end{bmatrix} \otimes \begin{bmatrix} 0 \\ 1 \end{bmatrix} = \begin{bmatrix} 0 \\ 0 \\ 0 \\ 0 \\ 0 \\ 1 \\ 0 \\ 0 \end{bmatrix} \tag{3}$$

Furthermore, qubits can become entangled, creating a profound interconnection between their states, even when physically separated, and facilitating more efficient calculations. The mathematical representation of a quantum register of size n is shown in Eq. (2). The symbol $\otimes$ is the tensorial product of two states (vectors). For example, the state of the quantum register $|\psi_{101}\rangle$ is described in Eq. (3).

A quantum gate is a fundamental element in quantum computing that manipulates the state of one or more qubits. Quantum gates are analogous to classical logic gates used in traditional computing. However, they operate on quantum states, which can exist in superposition and entangled states, allowing quantum computers to perform specific tasks much faster than classical computers. Quantum gates can operate on individual qubits or within quantum registers.

Some typical one-qubit quantum gates include the Pauli set, such as $X = \begin{bmatrix} 0 & 1 \\ 1 & 0 \end{bmatrix}$, which performs rotations on the X-axis, and $Y = \begin{bmatrix} 0 & -i \\ i & 0 \end{bmatrix}$, which performs rotations on the Y-axis. Another important one-qubit gate is the Hadamard gate $H = \frac{1}{\sqrt{2}} \begin{bmatrix} 1 & 1 \\ 1 & -1 \end{bmatrix}$ commonly used to put a qubit in a superposition state with probabilities of $50 - 50\%$ for the standard basis states $|0\rangle$ and $|1\rangle$, but with a difference in phase for the $|1\rangle$. A one-qubit quantum gate can operate on a quantum state $|\psi\rangle$ as follows $U|\psi\rangle$, where U is a quantum operator such as the quantum gate H. In this case, for the quantum state $|0\rangle$, we have the result $H|0\rangle = \frac{1}{\sqrt{2}} \begin{bmatrix} 1 & 1 \\ 1 & -1 \end{bmatrix} \begin{bmatrix} 1 \\ 0 \end{bmatrix} = \frac{1}{\sqrt{2}} \begin{bmatrix} 1 \\ 1 \end{bmatrix}$; i.e., $H|0\rangle = \frac{|0\rangle + |1\rangle}{\sqrt{2}}$.

Other more general one-qubit gates are the rotation gates, which are given by:

$$
\begin{aligned}
R_x(\theta) &= \begin{bmatrix} \cos\frac{\theta}{2} & -i\sin\frac{\theta}{2} \\ -i\sin\frac{\theta}{2} & \cos\frac{\theta}{2} \end{bmatrix} \\
R_y(\theta) &= \begin{bmatrix} \cos\frac{\theta}{2} & -\sin\frac{\theta}{2} \\ \sin\frac{\theta}{2} & \cos\frac{\theta}{2} \end{bmatrix} \\
R_z(\theta) &= \begin{bmatrix} e^{-\frac{i\theta}{2}} & 0 \\ 0 & e^{\frac{i\theta}{2}} \end{bmatrix}
\end{aligned} \tag{4}
$$

here, $R_x(\theta)$ denotes a rotation by an angle of θ radians about the x-axis, $R_y(\theta)$ represents a rotation of θ radians about the y-axis, and $R_z(\theta)$ signifies a rotation by θ radians around the z-axis As we mentioned before, a quantum register is composed of multiple qubits combined as a single entity. Similar to single quantum gates, in a quantum register, a quantum gate U_r can operate on the register state $|\psi_r\rangle^n$ of n-qubits as follows $U_r|\psi_r\rangle^n$. For example, in the simplest case of a $H^{\otimes 2}$ gate—two Hadamard gates in parallel, i.e., $H \otimes H$—operating on a two qubits register $|\psi_r\rangle^2$, we have $H^{\otimes 2}|\psi_r\rangle^2$, which is given by,

$$
H^{\otimes 2}|\psi_r\rangle^2 = \frac{1}{2} \begin{bmatrix} 1 & 1 & 1 & 1 \\ 1 & -1 & 1 & -1 \\ 1 & 1 & -1 & -1 \\ 1 & -1 & -1 & 1 \end{bmatrix} \begin{bmatrix} S_0 \\ S_1 \\ S_2 \\ S_3 \end{bmatrix} = \frac{1}{2} \begin{bmatrix} S_0' \\ S_1' \\ S_2' \\ S_3' \end{bmatrix} \tag{5}
$$

in this context, $S_0 \cdots S_3$ represent the complex quantum states of $|\psi_r\rangle^2$, and $S_0' \cdots S_3'$ denote the quantum state that results after the application of the quantum gate $H^{\otimes 2}$ to the state $|\psi_r\rangle^n$.

In general, for a state of n qubits $|0 \cdots 0_n\rangle$ in the computational basis, when we apply the Hadamard gate to each qubit, the resulting state is:

$$
|\psi_r\rangle^n = \frac{1}{\sqrt{2^n}}(|0 \cdots 000\rangle + |\cdots 001\rangle + \cdots + |0 \cdots 111\rangle) \tag{6}
$$

This transformation is very important since it is very common for many quantum algorithms to use it as an initial state because it can put into the superposition state with equal probabilities all the 2^n orthogonal qubits in the basis states $|0\rangle$ and $|1\rangle$. More generically, this transformation can be written as,

$$H^{\otimes n}|x\rangle^n = \frac{\sum_z (-1)^{x \odot z}|z\rangle}{\sqrt{2^n}} \tag{7}$$

where $x \odot z$ is the bitwise inner product of x and z.

Measurement plays a pivotal role within the framework of quantum mechanics, offering valuable insights into measuring physical observables by providing information about the possible measurement outcomes and their associated probabilities. It is essential to acknowledge that measurement disturbs a quantum system, leading to an irreversible change in its state. In quantum computation, measurement holds significant importance as it is crucial for retrieving information from the computational system.

In quantum mechanics, various measurement models are employed to clarify how a quantum system interacts with a measurement apparatus and how measurement outcomes are acquired. Some of the most common measurement models include the Projection Model, Expectation Value Model, State Collapse Model, Statistical Model, and Eigenvalue and Eigenspace Model.

Measurements can be described more broadly using standard notation. A measurement operator can be represented as M_m, where m serves as an index denoting a potential measurement outcome. When we have a state $|\psi\rangle$, the probability of obtaining measurement result m is as follows:

$$Pr(m) = \langle\psi|M_m^\dagger M_m|\psi\rangle \tag{8}$$

Once this measurement is obtained, the state immediately after this measurement will be:

$$|\psi'\rangle = \frac{M_m|\psi\rangle}{\sqrt{\langle\psi|M_m^\dagger M_m|\psi\rangle}} \tag{9}$$

The most common procedure in quantum computing involves measuring a group of qubits "in the standard basis $|0\rangle$ $|1\rangle$". This scenario, involves the act of examining the spin direction of individual qubits in the quantum memory register along an axis, in our case the z-axis according to the Bloch sphere. This z-axis is the one that passes through both the North and South poles of the sphere.

When performing these measurements, you will observe that each qubit is either in alignment with the z-axis, indicating "spin-up" (corresponding to state $|0\rangle$), or in opposition to it, signifying "spin-down" (corresponding to state $|1\rangle$). When applying this measurement to each qubit in an n-qubit quantum memory register, you'll end up with one of the 2^n possible bit string configurations that the register can assume.

The probability of achieving different results relies on the extent to which each binary string configuration exists in the superimposed state of the register right before measurement. For example, let us consider an n-qubit quantum memory register in the (normalized) state $\sum_{i=0}^{2^n-1} c_i|i\rangle$ (Note that the $|i\rangle$ represents a bit-string). The observed outcome will depend on the values of the amplitudes c_i and on whether you are performing a full measurement (measure all the qubits) or a partial measurement (measure only some qubits). Thus, on the standard basis, you will obtain the result $|i\rangle$ with a probability of $|c_i|^2$ for each state.

2.2 *Amazon Web Services*

AWS is a versatile platform that has grown and adapted to the needs of developers and emerging technologies. It offers a wide range of services, including cloud computing, storage, database processing, security, and advanced features such as high perfomance computing, machine learning and access to quantum computers.

The foundation of AWS is built on a network of companies and clients that use and provide software. Currently, AWS offers more than 300 services and has been in development for over 17 years. One of the services developed is Amazon Braket, which allows users to explore different quantum devices and companies available worldwide. It also has its own quantum simulator that utilizes AWS resources. Amazon Braket is a great alternative to start with quantum computing, distinct from other options like IBM Q, PennyLane, and Xanadu.

The name Amazon Braket comes from the standard "bra-ket" notation used in quantum mechanics. This platform offers base algorithms in quantum computing such as Grover, Bernstein-Vazirani, and Deutsch-Jozsa, among others, all of which are written in Python. Additionally, Amazon Braket allows us to store all the results of executions and run the algorithms without needing individual access to a specific provider of these devices, making it a convenient and user-friendly platform for quantum computing.

In addition to the quantum devices listed in Table 1, it also provides access to three simulators, SV1, DM1, and TN1, in which SV1 is a tensor network with 50 qubits, DM1 is a vector state simulator with 34 qubits, and TN1 uses a density matrix with 17 qubits.

Quantum gates are applied to qubits through pulses that are sent to them. These pulses form a wave that is driven by a signal that is connected to an input port on the hardware. The signal comprises specific parameters such as amplitude, frequency, and duration.

Once a quantum circuit has been executed, it runs 100 times to determine which result will most likely be obtained. This method is used to obtain the measurement. Although this parameter can be customized to suit your needs, it is recommended that there be a minimum of 50 runs.

A crucial component for storing the outcomes of our operations is "Amazon S3 bucket" (Amazon Simple Storage Service). The service allows data to be saved as

Table 1 Available quantum devices on amazon braket

Companies	Quantum device names	Quantum computing technologies
IonQ[1]	Aria 1, Harmony	It uses precise laser pulses, monitoring ions trapped in space to perform quantum gate operations and measurements. At the moment there are 11 and 25 qubits available
Oxford Quantum Circuits (OQC)[2] and Rigetti[3]	Lucy, Aspen M-3	Its technology uses Superconducting Quantum Processors and has 8 qubits, while Reggeti is made up of two chips with in total of 79 qubits
QuEra computing[4]	Aquila	In this technology, lasers are used to arrange and excite neutral atoms into highly energetic states. They have a quantum computer with 256 qubits in analog mode. His paradigm is with analog Hamiltonian Simulation

The only device with support for MATLAB (at the moment) is Harmony
[1] https://ionq.com/quantum-systems/aria
[2] https://qcs.rigetti.com/qpus
[3] https://oxfordquantumcircuits.com/oqc-on-aws
[4] https://www.quera.com/

objects in a bucket, providing a maximum storage capacity of 5 TB. It is possible to store any type of file, including videos, text files, and Braket task results.

A task is defined in "Amazon Braket" and then sent to the device for execution, such as a quantum circuit or a quantum register. In some cases, the task is queued until the device is ready before being sent to the quantum device or simulator. When jobs are submitted to a quantum device, they are processed by external companies that operate facilities with available quantum computers.

After the task is completed, "Braket" will send the outcomes to a secure bucket created in S3. You can easily monitor and manage all tasks on the "Quantum Tasks" page through the "Amazon Braket console".

3 Quantum Genetic Algorithm

An evolutionary algorithm is an optimization method inspired by biological evolution used to find efficient or approximate solutions to complex problems across various disciplines. These algorithms initiate with a population of potential solutions, evaluate each one's fitness using a predefined function, select individuals for reproduction based on fitness, generate new solutions through crossover and mutation, and repeat this process over multiple generations. The primary objective is to discover optimal or near-optimal solutions in intricate search spaces, making evolutionary algorithms invaluable for tackling real-world problems in areas like engineering, computer science, and artificial intelligence.

Our work focuses on QGAs, which are rooted in Quantum-Inspired Genetic Algorithms (QIGAs) [8]. The two categories of algorithms involve quantum principles but differ in their implementation and computational framework. The QIGAs are classical algorithms executed on conventional computers, but quantum principles inspire them and aim to mimic quantum effects through classical computing techniques. QGAs, on the other hand, are designed to run on true quantum computers, taking full advantage of quantum properties. They have the potential for significant speedup for certain problems, and currently are limited by the nascent state of quantum hardware and are not widely accessible for practical use.

The specific approach of this study involves implementing a quantum genetic algorithm in MATLAB to be executed in a quantum computer through the Amazon Braket. We use the "quantum library of MATLAB" [7, 9, 10]. We illustrate the methodology with a straightforward example.

3.1 Proposal

We utilize an array of 11 qubits to represent a single quantum chromosome. The population is composed of 10 individuals. A quantum chromosome, denoted as q_i, is defined as a sequence of m qubits, where $i = 10$ and $m = 11$. For simplicity, we will use the notation q_i. In this context, m represents the genetic makeup of each individual, and i signifies the population size, characterizing a quantum system $|\psi_i\rangle$ that encompasses 2^m simultaneous states, as expressed in Eq. (10).

$$|\psi_i\rangle = \begin{bmatrix} \alpha_1 \ \alpha_2 \ \ldots \ \alpha_m \\ \beta_1 \ \beta_2 \ \ldots \ \beta_m \end{bmatrix} = q_i \tag{10}$$

The result will yield a quantum population in the following structure:

$$\begin{bmatrix} \alpha_1 \ \alpha_2 \ \ldots \ \alpha_m \\ \beta_1 \ \beta_2 \ \ldots \ \beta_m \end{bmatrix}_{q_1}$$

$$\begin{bmatrix} \alpha_1 \ \alpha_2 \ \ldots \ \alpha_m \\ \beta_1 \ \beta_2 \ \ldots \ \beta_m \end{bmatrix}_{q_2}$$

$$\vdots$$

$$\begin{bmatrix} \alpha_1 \ \alpha_2 \ \ldots \ \alpha_m \\ \beta_1 \ \beta_2 \ \ldots \ \beta_m \end{bmatrix}_{q_i}$$

Our implementation is founded on QIGA [11]. We have customized this approach to enable its execution on a quantum computing platform, as discussed in [7]. As a result, Algorithm 1 and Algorithm 2 include these adaptations. Algorithm 1 requires two input parameters: the number of generations, g, for the population to evolve and the population size s_p. As mentioned earlier, the population and quantum chromosome sizes have been fixed. In line 2, the generation counter t is initialized in 0.

In line 3, to ensure equal probabilities for all potential individuals, we applied the quantum Hadamard gate to all the qubits, as defined in Eq. (7).

In line 4, we obtain a classical population by measuring all individuals in the current population. In line 5, we store in "**b**" the best solution in the current population. The while-loop covering lines 6 to 10 is executed while the generation counter is smaller than the maximum number of generations "g". In step 7, we generate a new quantum population using the best solution (individual) "**b**" and update it using Algorithm 2. In step 8, a new classical population is created by taking measurements of the individuals obtained in step 7. In step 9, we upgrade the generation counter. In the end, the best solution of the whole process is stored in "**b**".

Algorithm 1 Quantum Genetic Algorithm (QGA)

Require: g, s_p ▷ g: number of generations, s_p: population size
Ensure: The best solution **b**
1: **procedure** QGA(g, s_p)
2: $t \leftarrow 0$
3: Initialize a quantum population $Q_t(s_p)$ with quantum gate H
4: Make $P_t(s_p)$, measure of each individual $Q_t(s_p) \rightarrow P_t(s_p)$
5: Store the best solution **b** in $P_t(s_p)$
6: **while** $t <$ g **do**
7: **Update** $Q_t(s_{p-1})$ using a quantum rotation gate $U(\Delta\theta_j)$
8: **Make** $P_{t+1}(s_{p-1})$, measure of each individual $Q_{t+1}(s_{p-1}) \rightarrow P_t(s_{p-1})$
9: $t \leftarrow t+1$
10: **end while**
11: **end procedure**

Algorithm 2 Update of the QGA

Require: q_i, x_i, b_i ▷ q_i quantum element, x_i and b_i the best element decimal and binary respectively ▷ Also we need $f(b_i)$ and $f(x_i)$
Ensure: Evolution rate
procedure UPDATE(q_i)
$i \leftarrow 0$
while $i < m$ **do** ▷ m is the size of q_i
Determine $\Delta\theta_i$ and $s(\alpha_i', \beta_i')$ with help to Table 2
Update q_i applying :
$[\alpha_i' \beta_i']^T = U(\Delta\theta_i)[\alpha_i, \beta_i]^T$
end while
$q_i = q_i'$
end procedure

Algorithm 2 provides the required rotation angle that will eventually modify the amplitude probability to change individual qubits of the quantum chromosomes. The algorithm is based on Table 2 to choose the required angle to use Eq. (11) that will update the quantum state of the corresponding qubit in the quantum chromosome. The original rotation angles proposed in [11] were modified to those shown in Table 2

Table 2 Values and conditions for rotation $\Delta\theta_i$

x_{ij}	b_{ij}	$f(X) \geq f(b_i)$	$\Delta\theta_i$
0	0	*False*	0
0	0	*True*	0
0	1	*False*	0
0	1	*True*	-0.05π
1	0	*False*	-0.01π
1	0	*True*	0.025π
1	1	*False*	0.005π
1	1	*True*	0.025π

Source This table was adapted in according to [7, 9]

specifically for this algorithm.

$$\left[\alpha'_x \beta'_x\right] = \begin{bmatrix} \cos(\Delta\theta_i) & -\sin(\Delta\theta_i) \\ \sin(\Delta\theta_i) & \cos(\Delta\theta_i) \end{bmatrix} \begin{bmatrix} \alpha_x \\ \beta_x \end{bmatrix} \quad (11)$$

The integration between MATLAB and a specific quantum device, detailed in Table 1, is facilitated by a newly developed library named "quantum", which was introduced in the current year, 2023. The connection process is initiated through the command demonstrated in Code 1. It is worth noting that to establish this connection, an Amazon Web Service account is a prerequisite. The `dev` variable represents the selected quantum device, while the `task` variable pertains to the quantum circuit design.

Here are the necessary steps to establish the connection:

1. Begin by downloading and installing the "quantum" library. Please note that this library is compatible with MATLAB versions starting from 2023a.
2. Create an AWS account of the IAM type, which is designed for daily task execution. This process includes:
   (a) Creating another user and generating an "access key" from the Summary section.
   (b) Granting permissions for access to Amazon Braket services.
   (c) Providing permissions for access to AWS S3, a cloud-based storage solution.
3. Save the access credentials in a file with the extension ".env". With the following form:
   (a) `AWS_ACCESS_KEY_ID=**ACCESS IDENTIFIER PROVIDED**`
   (b) `AWS_SECRET_ACCESSS_KEY=**GENERATED ACCESS KEY**`
   (c) `AWS_DEFAULT_REGION=**REGION OF THE SELECTED QUANTUM DEVICE**`
4. Utilize the `loadenv` command to load the information from the "file.env".
5. Create an object "Bucket" named so that it starts with `amazon-braket-`.

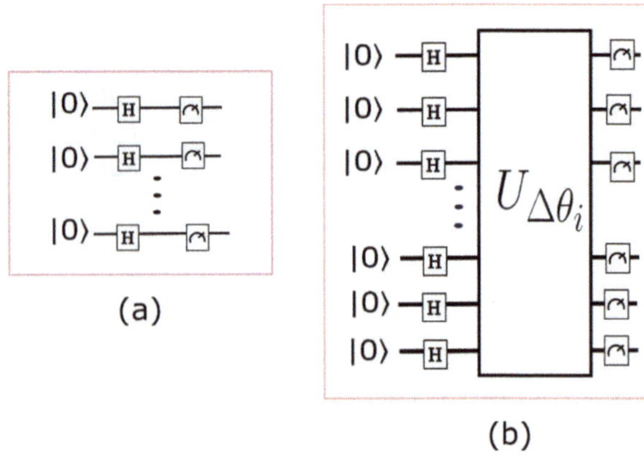

Fig. 1 Operators of quantum genetic algorithm. **a** Diagram of the quantum circuit for the initialization of one population. **b** Diagram corresponding to a quantum circuit for the mutation and crossover operators

6. Inside this bucket, create a folder to store the results of the executions.
7. Get "Copy S3 URL" from the folder created in the bucket. This will be our "bucketPath".
8. Verify that the region settings for both the bucket and the quantum device are consistent.
9. Explore the available AWS quantum devices and simulators.
10. Selection one device and copy "Device ARN", this information should be added in the Code 1 on the variable "devARN".
11. Use Code 1 to start the connection.

Code 1 Connection to quantum device with MATLAB

```
loadenv('awsConfig.env')
reg = "us-east-1";
bucketPath = "s3://amazon-braket-nameBucket/nameFolder/";
devARN="arn:aws:braket:us-east-1::device/qpu/ionq/Harmony";
dev = quantum.backend.QuantumDeviceAWS(devARN,Region=reg,S3Path=bucketPath);
qc = quantumCircuit(HGate(1));        %Example of quantum circuit
task = run(qc,dev);
wait(task);
m = fetchOutput(task);
%The next line shows the results in a table from
table(m.Counts,m.MeasuredStates, VariableNames=["counts","States"])
```

Figure 1 shows the quantum circuits used by the quantum genetic algorithm; for the initial quantum population and by the process for mutation and crossover. Their pseudo-codes are shown in Algorithm 1 and Algorithm 2.

4 Experiments and Results

In this section, we show how to implement the quantum genetic algorithm on a simple optimization problem using the library "quantum" from MATLAB to run

Table 3 Results on six continuous functions with 15 generations

No.	Function ∈ [0, 127]	x_{max}	$f(x_{max})$	Time (s)
1.	$f(x) = \frac{x}{10} \cdot sin(\frac{x}{10})$	80.0625	7.9	3893.4884
2.	$f(x) = \frac{x}{2} + 5$	124	67	112.82994
3.	$f(x) = -\frac{cos(x) \cdot x^2}{10+3x}$	116.1875	1696.7312	93.799
4.	$f(x) = \frac{(x-2\pi 65)(sen(x+2\pi 65))}{10}$	15.4375	39.2826	519.50.20
5.	$f(x) = \frac{-x^2}{10+3x}$	15	22.5	944.2783
6.	$f(x) = \frac{sin(x-20)}{x-20}$	20.0625	0.99935	1005.5055

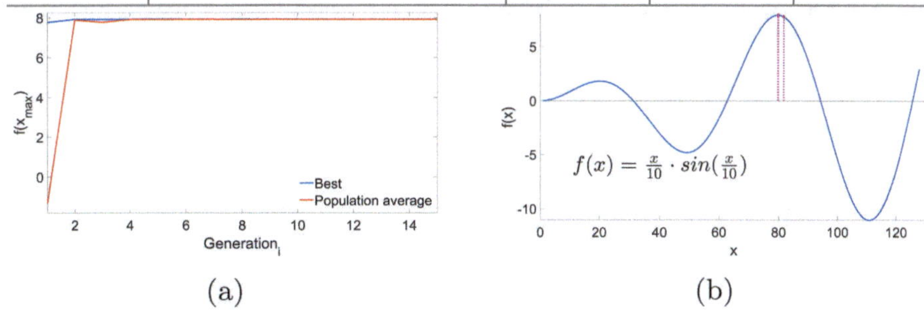

Fig. 2 **a** Population's average and the behavior of the best values found. **b** maximum value for function 1 found after one execution with 15 generations

in an available quantum device of AWS. This optimization aims to find the value x, which maximizes the value of the functions shown in Table 3. The numbers were taken in binary and transformed into decimal values, as shown in the previous section. Each experiment was executed for 15 generations.

To execute the MATLAB program on the AWS Braket platform,[1] whether in the simulator or on the quantum device, it is necessary to create a key for remote connection and follow the instructions outlined in the preceding subsection (see Sect. 3.1) as detailed in Code 1. After these setup steps, the chosen device for running the quantum circuits designed (Fig. 1) for each quantum genetic operator in the QGA was "Harmony", which was selected because it was the only available device with support in MATLAB, offering a total of 11 available qubits.

About the behavior of the Quantum Genetic Algorithm (QGA), the results depicted in Figs. 2, 3, 4, 5, 6, and 7 illustrating that the QGA requires only a few generations to reach optimal or near-optimal values. The configuration of each figure mentioned is as follows, (a) shows an average that depicts how the first experiments found the values near to optimum even the optimum value, and (b) shows that the numbers of "good" results are few.

[1] https://aws.amazon.com/es/blogs/quantum-computing/setting-up-your-local-development-environment-in-amazon-braket/.

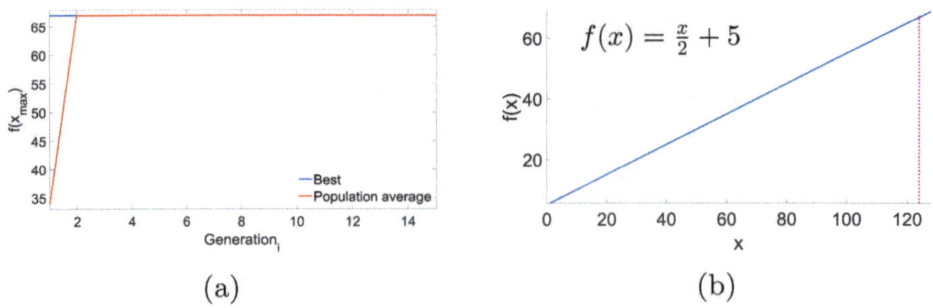

Fig. 3 **a** Population's average and the behavior of the best values found. **b** maximum value for function 2 found after one execution with 15 generations

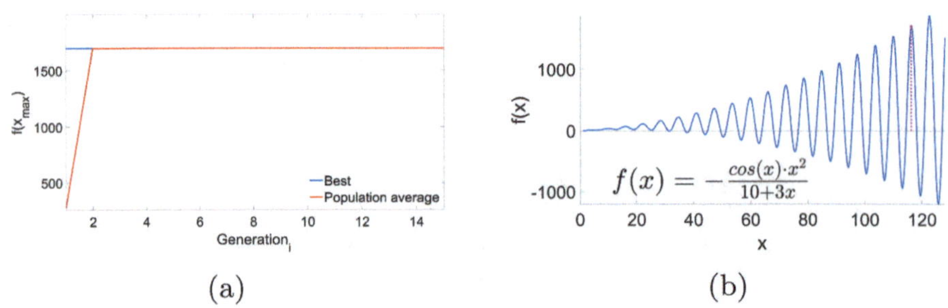

Fig. 4 **a** Population's average and the behavior of the best values found. **b** maximum value for function 3 found after one execution with 15 generations

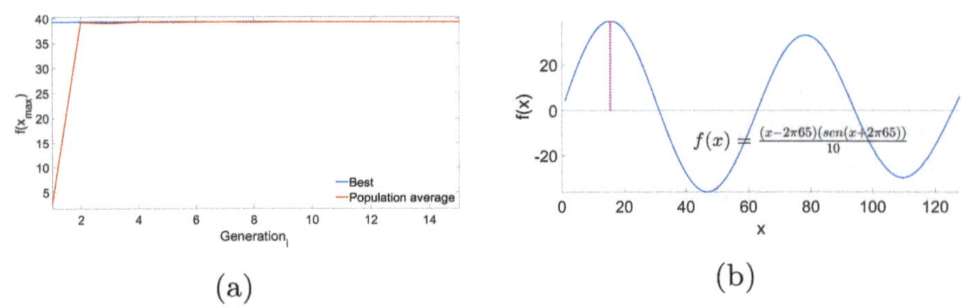

Fig. 5 **a** Population's average and the behavior of the best values found. **b** maximum value for Function 4 found after 10 executions with 15 generations run on the "Amazon SV1" simulator

This is a promising outcome, especially when compared to simulation results from previous studies such as [9] and [12], which also demonstrated favorable results. One noteworthy distinction in our work is the time factor, as accessing AWS can be time-consuming and contingent upon internet availability.

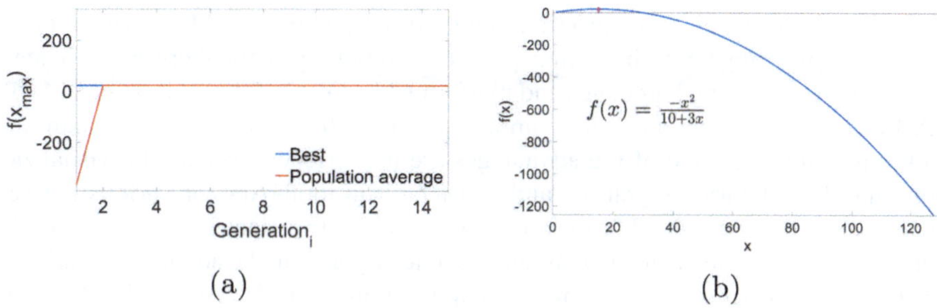

Fig. 6 **a** Population's average and the behavior of the best values found. **b** maximum value for function 5 found after 10 executions with 15 generations run on the "Amazon SV1" simulator

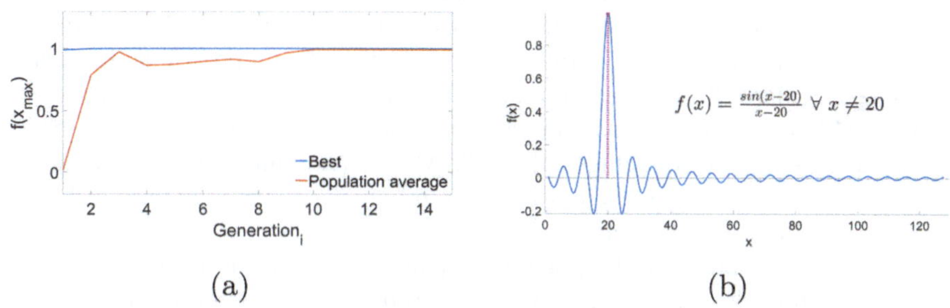

Fig. 7 **a** Population's average and the behavior of the best values found. **b** maximum value for function 6 found after 10 executions with 15 generations run on the "Amazon SV1" simulator

5 Conclusion

Quantum genetic algorithms offer a promising approach to tackling high-dimensional complex problems that cannot be solved efficiently by classical computing methods, including heuristics, metaheuristics, linear and integer programming, dynamic programming, and other conventional techniques. In essence, quantum approaches harness the inherent characteristics of quantum physics, such as superposition and entanglement, thereby endowing these methods with exponential capabilities to reduce execution time and resource utilization.

Quantum computing is currently in the NISQ era, which brings several significant challenges and problems, primarily due to the limitations of current quantum hardware. Some of the key challenges in this era include quantum errors and noise that significantly impact the accuracy of quantum computations, the restricted depth of circuits that limits the number of quantum operations before cumulative errors become problematic, the limited number of available qubits, restricted access to quantum devices, and hardware variability, among others. These challenges necessitate hybrid computation by combining classical and quantum computing, introducing complexity, and reducing the speedup potential.

MATLAB is a widely used programming and computational tool in various scientific and engineering fields, including quantum computing. While there are dedicated quantum programming languages and platforms like Qiskit, Cirq, and others, MATLAB has advantages in quantum computing, especially for those already familiar with this software; some of the advantages are user-friendly, versatility, visualization, interdisciplinary integration, and many built-in toolboxes for various mathematical and numeric tasks. For the above reasons, we investigated through a simple optimization example, using a Quantum Genetic Algorithm, the advantages and disadvantages of programming quantum computing through AWS using the Amazon Bracket.

An existing limitation within the context of available quantum devices for MATLAB is the restricted support provided for MATLAB on the AWS Braket portal. While the portal claims the availability of six quantum devices, it is noteworthy that only "Harmony" offers compatibility with MATLAB, further constrained by its limited capacity of 11 qubits. This limitation, in turn, restricts the scope of practical applications, particularly those reliant on a hybrid scheme, as significant latency exists between MATLAB and the quantum device. Despite these constraints, it is imperative to underscore the advantages of MATLAB for quantum device programming. The factors above allude to the considerable promise associated with MATLAB's quantum programming capabilities.

It is exciting to envision the potential expansion of Amazon Braket's quantum device portfolio, which may encompass more robust devices featuring larger qubit arrays and diminished noise levels. These advancements must be poised to expedite quantum computations significantly. Moreover, the identified limitations discussed here promise future refinement, rendering quantum computing increasingly feasible for addressing complex problems that classical computing has yet to tackle thoroughly. In essence, some of the ongoing research relying on classical computing techniques, such as metaheuristic algorithms, might transition towards the realm of quantum computing, offering the prospect of enhancing quantum genetic algorithms, among other possibilities.

Acknowledgements This work was supported by Instituto Politécnico Nacional (institutional project SAPI (20230104)), and by the Mexican National Council of Science and Technology (CONAHCYT, México) project CF-2023-I-108. We additionally thank to the Coordinación Institucional de Investigación of CETYS Universidad for their collaboration with the work, and to RedCIBaja-AWS.

References

1. M. Mitchell, *An Introduction to Genetic Algorithms* (Bradford, Cambridge, Massachusetts, USA, 2001)
2. R.L. Haupt, S.E. Haupt, *Practical Genetic Algorithms* (Wiley, Hoboken, New Jersey, USA, 2004)

3. S.S. Gill, A. Kumar, H. Singh, M. Singh, K. Kaur, M. Usman, R. Buyya, Quantum computing: a taxonomy, systematic review and future directions. Softw.: Practice Exp. **52**(1), 66–114 (2022). https://doi.org/10.1002/spe.3039
4. J.R. Wootton, F. Harkins, N.T. Bronn, A.C. Vazquez, A. Phan, A.T. Asfaw, Teaching quantum computing with an interactive textbook, in *2021 IEEE International Conference on Quantum Computing and Engineering (QCE)* (2021), pp. 385–391. https://doi.org/10.1109/QCE52317.2021.00058
5. M. AbuGhanem, H. Eleuch, Experimental characterization of google's sycamore quantum AI on an IBM's quantum computer. Available at SSRN 4299338 (2023). https://doi.org/10.2139/ssrn.4299338
6. P.J. Karalekas, N.A. Tezak, E.C. Peterson, C.A. Ryan, M.P. da Silva, R.S. Smith, A quantum-classical cloud platform optimized for variational hybrid algorithms. IOP Publishing (2020). https://doi.org/10.1088/2058-9565/ab7559
7. O.H. Montiel Ross, A review of quantum-inspired metaheuristics: Going from classical computers to real quantum computers. IEEE Access **8**, 814–838 (2020). https://doi.org/10.1109/ACCESS.2019.2962155
8. K.-H. Han, J.-H. Kim, Genetic quantum algorithm and its application to combinatorial optimization problem, in *Proceedings of the 2000 Congress on Evolutionary Computation. CEC00 (Cat. No.00TH8512)*, vol. 2 (2000), pp. 1354–13602. https://doi.org/10.1109/CEC.2000.870809
9. R. Lahoz-Beltra, Quantum genetic algorithms for computer scientists. Computers **5**(4) (2016). https://doi.org/10.3390/computers5040024
10. E. Ballinas, O. Montiel, Hybrid quantum genetic algorithm with adaptive rotation angle for the 0–1 knapsack problem in the ibm qiskit simulator. Soft Comput. **27**(18), 13321–13346 (2023). https://doi.org/10.1007/s00500-022-07460-7
11. K.-H. Han, J.-H. Kim, Quantum-inspired evolutionary algorithm for a class of combinatorial optimization. IEEE Trans. Evol. Comput. **6**(6), 580–593 (2002). https://doi.org/10.1109/TEVC.2002.804320
12. G. Acampora, A. Chiatto, A. Vitiello, Genetic algorithms as classical optimizer for the quantum approximate optimization algorithm. Appl. Soft Comput. **142**, 110296 (2023). https://doi.org/10.1016/j.asoc.2023.110296

Intensification Process Applied to Codon Selection in Grammatical Evolution

Juan Manuel Perez-Rocha, J. Martín Carpio, Alfonso Rojas-Domínguez, Manuel Ornelas-Rodríguez, Héctor J. Puga-Soberanes, and Marco Aurelio Sotelo-Figueroa

1 Introduction

Automatic Programming (AP) [1–3] and Genetic Programming (GP) [4, 5] are two prominent approaches in the field of evolutionary computation that aim to automatically generate computer programs to solve various tasks. These techniques have been instrumental in solving complex problems and optimizing algorithms. However, they each come with their own set of limitations and challenges. To address some of these limitations, Grammatical Evolution (GE) [6] emerged as a promising evolution-inspired technique that joins the power of GP with a more structured approach based on formal grammar.

GP is an evolutionary algorithm that evolves computer programs using a population of candidate solutions, represented as tree structures. Through the application of mutation, crossover, and selection operators, GP aims to evolve programs that are capable of performing a specific task. While powerful, GP often faces issues related to program bloat, inefficiency, and a lack of control over the search space [7].

GE provides a solution to these challenges by introducing a mapping process that bridges the gap between the genotype and phenotype of an evolved program. It does this by employing formal context-free grammars [8]. Instead of directly evolving program trees, Grammatical Evolution evolves strings of symbols based on a specified grammar. These strings are then mapped to valid programs, thus ensuring the syntactic and semantic correctness of the generated solutions.

One of the key advantages of GE is the separation of genotype (the evolving string) from phenotype (the resulting program) [9]. This separation allows for better control

J. M. Perez-Rocha · J. M. Carpio (✉) · A. Rojas-Domínguez · M. Ornelas-Rodríguez · H. J. Puga-Soberanes
División de Estudios de Posgrado e Investigación, Tecnológico Nacional de México—Campus León, Av. Tecnológico S/N, León 37290, Guanajuato, Mexico
e-mail: juanmartin.carpio@leon.tecnm.mx

M. A. Sotelo-Figueroa (✉)
Departamento de Estudios Organizacionales, Universidad de Guanajuato, Fraccionamiento I El Establo, Guanajuato 36250, Guanajuato, Mexico
e-mail: masotelo@ugto.mx

P. Melin and O. Castillo (eds.), *New Directions on Hybrid Intelligent Systems Based on Neural Networks, Fuzzy Logic, and Optimization Algorithms*, Studies in Computational Intelligence 1146, https://doi.org/10.1007/978-3-031-53713-4_11

over the search space, as the grammar can be designed to enforce certain structural and syntactical constraints, reducing the likelihood of generating non-functional or inefficient programs. Additionally, the search space can be systematically explored by varying the genotype, providing a more systematic approach to program generation.

Most of the GE research is based on search engines, including various metaheuristics such as Genetic Algorithm (GA) [10], Differential Evolution (DE) [11], Particle Swarm Optimization (PSO) [12], Estimation Distribution Algorithms (EDA) [13] among others. However, in [14] it is proposed to work with the order of the codons as a search space instead of the values of the codons.

In the context of symbolic regression [15, 16], GE has found widespread application. Symbolic regression is a problem where the goal is to discover mathematical expressions or functions that best fit a given dataset. By applying Grammatical Evolution, a formal grammar can be defined that generates mathematical expressions and functions, ensuring they are syntactically valid. The search engine within GE explores the space of possible expressions, allowing it to discover highly accurate models to represent the data.

This paper applied a 2-opt heuristic to make a local search over the codon permutation order to guide the search on the Grammatical Evolution. It was applied to the Symbolic Regression Problem instances, the results were compared statistically with classic GE and random permutations GE using a non-parametric test.

2 Grammatical Evolution

Grammatical Evolution (GE) is a computational technique that blends elements of genetic programming and formal grammars to evolve computer programs or solutions to various problems.

Figure 1 shows the methodology used in GE. It shows the search engine, BNF Grammar, Mapping Process, and the Problem Instance as the main GE components.

Figure 2 shows a classical derivation example using the Depth-First [18] mapping process and a Backus Naur Form (BNF) grammar [19]. The codons are marked as gen.

3 Symbolic Regression Problem

The Symbolic Regression Problem (SRP) [15, 16, 21] is a computational task in the field of machine learning and evolutionary computation, where the goal is to discover a mathematical expression or function that best fits a given dataset. In symbolic regression, the algorithm seeks to find a symbolic representation (an equation or a mathematical formula) that describes the relationship between the input variables and the target output variable, without prior knowledge of the functional form.

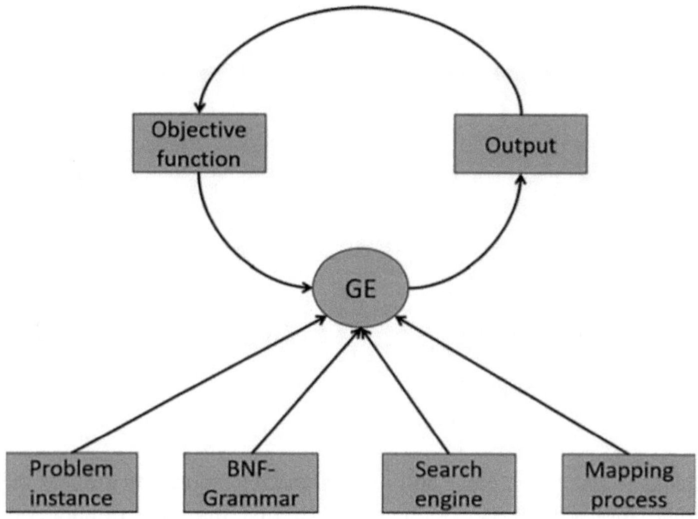

Fig. 1 GE methodology [17]

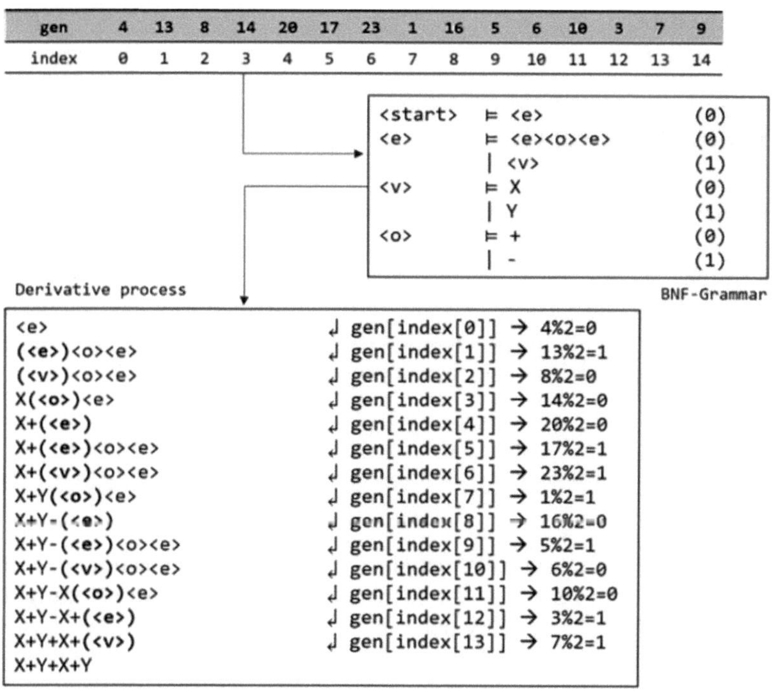

Fig. 2 Classical derivation example [20]

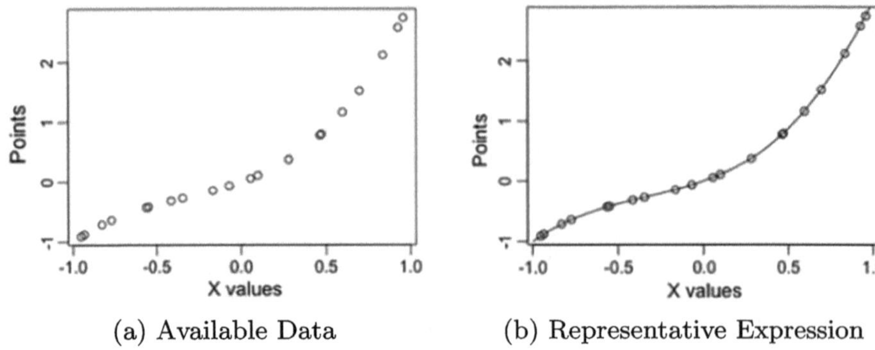

Fig. 3 SRP based on the equation $X^3 + X^2 + X$

The goal is to provide a clear and understandable representation of the relationships within complex datasets, which can lead to insights and actionable information.

Figure 3 shows an example of SRP. Figure 3a shows the available data that can be obtained without knowing the function and Fig. 3b shows the expression that generates the data.

4 Proposed Approach

Figure 4 shows the proposal from [14] in which the codons are not taken sequentially but are taken in the order shown by per. Per is an array generated randomly based on a permutation.

If the fitness does not get a better value during iterations, it applies a 2-opt heuristic (as shown in Sect. 4.1) to the permutation index to choose the codons. Algorithm 1 shows the GE algorithm with the current proposal.

4.1 2-Opt

The 2-opt heuristic [22, 23] is considered a local search technique because it focuses on making small adjustments to the current solution to improve it locally. It does not guarantee finding the globally optimal solution but is effective in improving an initial solution quickly. The 2-opt heuristic is an approach to improve the quality of a solution by iteratively swapping pairs of items.

Figure 5 shows an example of 2-opt.

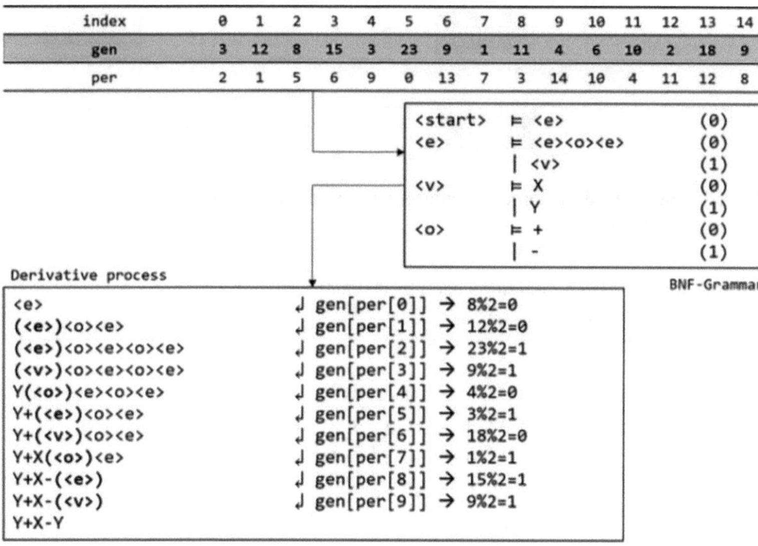

Fig. 4 Derivation example based on permutations [20]

Algorithm 1 Grammatical Evolution Algorithm

Require: $pop\_size, dim, search\_engine, cont\_apply\_LS$

$Population \leftarrow new\_population(pop\_size, dim)$
$Per \leftarrow generate\_permutation(dim)$
$Fitness \leftarrow Evaluate(Population, Per)$
$Cont \leftarrow 0$
$GBest \leftarrow get\_Best(Population, Fitness)$
while termination condition not met **do**
  $Population \leftarrow search\_engine(Population)$
  $Fitness \leftarrow Evaluate(Population, Per)$
  $Best \leftarrow get\_Best(Population, Fitness)$
  if $GBest = Best$ **then**
    $cont\_apply\_LS \leftarrow cont\_apply\_LS + 1$
  end if
  if $Cont = cont\_apply\_LS$ **then**
    $Per \leftarrow 2opt(Per)$
    $Cont \leftarrow 0$
  end if
end while
return $BestSolution$

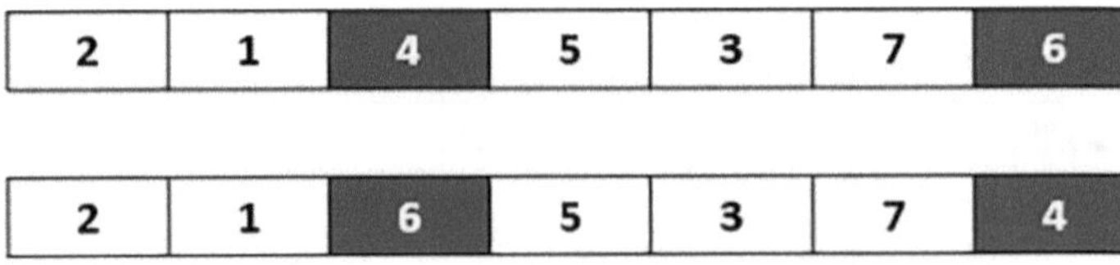

Fig. 5 2-opt example, the first element contains a permutation from 1 to 7, was chosen two elements and they were interchanged

Table 1 Symbolic regression functions used as instances set

Function	Polynomial
F_1	$f(x) = X^3 + X^2 + X$
F_2	$f(x) = X^4 + X^3 + X^2 + X$
F_3	$f(x) = X^5 + X^4 + X^3 + X^2 + X$
F_4	$f(x) = X^6 + X^5 + X^4 + X^3 + X^2 + X$
F_5	$f(x) = sin\left(x^2\right) cos(x) - 1$
F_6	$f(x) = sin(x) + sin\left(x + x^2\right)$
F_7	$f(x) = log(x+1) + log\left(x^2+1\right)$
F_8	$f(x) = \sqrt{x}$
F_9	$f(x, y) = sin(x) + sin\left(y^2\right)$
F_{10}	$f(x, y) = 2sin(x) cos(y)$
$Keijzer_1$	$f(x) = 0.3xsin(2\pi x)$
$Keijzer_2$	$f(x) = 1 + 3x + 3x^2 + x^3$
$Keijzer_3$	$f(x, y) = 8/(2 + x^2 + y^2)$
$Keijzer_4$	$f(x, y) = x^4 - x^3 + y^2/2 - y$
$Keijzer_5$	$f(x, y) = x^3/5 + y^3/2 - y - x$
$Keijzer_6$	$f(x_1, x_2, \ldots, x_{10}) =$ $10.59x_1x_2 + 100.5967x_3x_4 - 50.59x_5x_6 + 20x_1x_7x_9 + 5x_3x_6x_{10}$

5 Experimental Setup

The functions used are shown in Table 1, the functions F_1 to F_{10} were taken from [20, 21] and the Keijzer from [24–26].

The Grammars used are the following:

- Grammar 1 for functions F_1 to F_8.
- Grammar 2 for functions F_9 to F_{10}.
- Grammar 3 for functions $Keijzer_1$ to $Keijzer_5$.
- Grammar 4 for function $Keijzer_6$.

To discern the quality of the expressions given by the GE, the Mean Root Squared Error (MRSE) given by Eq. 1.

$$
\begin{aligned}
\langle \text{start} \rangle &\models \langle \text{expr} \rangle \\
\langle \text{expr} \rangle &\models (\langle \text{expr} \rangle \langle \text{op} \rangle \langle \text{expr} \rangle) \mid \langle \text{pre-op} \rangle (\langle \text{expr} \rangle) \mid \langle \text{var} \rangle \\
\langle \text{var} \rangle &\models \texttt{x} \mid \texttt{1} \\
\langle \text{pre-op} \rangle &\models \texttt{sin} \mid \texttt{cos} \mid \texttt{exp} \mid \texttt{log} \\
\langle \text{op} \rangle &\models + \mid - \mid / \mid *
\end{aligned}
$$

Grammar 1 Grammar for the functions F_1 to F_8

$$
\begin{aligned}
\langle \text{start} \rangle &\models \langle \text{expr} \rangle \\
\langle \text{expr} \rangle &\models (\langle \text{expr} \rangle \langle \text{op} \rangle \langle \text{expr} \rangle) \mid \langle \text{pre-op} \rangle (\langle \text{expr} \rangle) \mid \langle \text{var} \rangle \\
\langle \text{var} \rangle &\models \texttt{x} \mid \texttt{y} \mid \texttt{1} \\
\langle \text{pre-op} \rangle &\models \texttt{sin} \mid \texttt{cos} \mid \texttt{exp} \mid \texttt{log} \\
\langle \text{op} \rangle &\models + \mid - \mid * \mid /
\end{aligned}
$$

Grammar 2 Grammar for the functions F_9 to F_{10}

$$
\begin{aligned}
\langle \text{start} \rangle &\models \langle \text{expr} \rangle \\
\langle \text{expr} \rangle &\models (\langle \text{expr} \rangle \langle \text{op} \rangle \langle \text{expr} \rangle) \mid \langle \text{pre-op} \rangle (\langle \text{expr} \rangle) \mid \langle \text{var} \rangle \\
\langle \text{var} \rangle &\models \texttt{x} \mid \texttt{y} \mid \texttt{1} \\
\langle \text{op} \rangle &\models + \mid - \mid * \mid /
\end{aligned}
$$

Grammar 3 Grammar for the functions Keijzer from 1 to 5

$$
\begin{aligned}
\langle \text{start} \rangle &\models \langle \text{expr} \rangle \\
\langle \text{expr} \rangle &\models (\langle \text{expr} \rangle \langle \text{op} \rangle \langle \text{expr} \rangle) \mid \langle \text{pre-op} \rangle (\langle \text{expr} \rangle) \mid \langle \text{var} \rangle \\
\langle \text{var} \rangle &\models \texttt{x}\langle \text{c} \rangle \mid \texttt{1} \\
\langle \text{op} \rangle &\models + \mid - \mid * \mid / \\
\langle \text{c} \rangle &\models 0 \mid 1 \mid 2 \mid 3 \mid 4 \mid 5 \mid 6 \mid 7 \mid 8 \mid 9
\end{aligned}
$$

Grammar 4 Grammar for the function Keijze 6

$$
MRSE = \sqrt{\frac{1}{n}\sum_{i=1}^{n}(t_i - y_i)^2} \tag{1}
$$

Table 2 GA parameters

Parameter	Value
Population size	300
Dimensions	100
Iterations to apply 2-opt	5
Function evaluations	250,000
Selection	Binary tournament
Crossover	2-points
Mutation	Bit-flip

where:

- n is the number of data points.
- y_i is the real value.
- t_i corresponds to the value obtained.

The classic GE, Random Permutation GE, and the proposal use GA as a Search Engine, the parameters used are shown in Table 2.

For each function, it was performed 33 experiments and was compared using the median to compare the results. The results were compared with those obtained by the classical GE and the Random permutation GE. The comparison was implemented using the Friedman non-parametric test to discern if there exists some implementation that performs better.

6 Results

Table 3 shows the results obtained for each instance. The results shown are the medians of the 33 experiments. With these data, a Friedman non-parametric test was performed. The value obtained with this test was 2.375 with a p-value 0.304982. Due to the p-value being bigger than 0.1 is not possible to perform a post-hoc procedure to determine whether some variant of GE gives better results.

Table 4 shows the results obtained for each instance. The results shown are the standard deviation of the 33 experiments. With these data, a Friedman non-parametric test was performed. The value obtained with this test was 13.03125 with a p-value 0.001480131. Due to the p-value being smaller than 0.1 is possible to perform a post-hoc procedure to determine whether some variant of GE gives better results, the ranking is shown in Table 5.

Table 3 Median of the results obtained for each instance and GE variant

Instance	Classical GE	Random permutations	Proposal
F_1	0.026	0.015	0.018
F_2	0.023	0.018	0.023
F_3	0.025	0.025	0.038
F_4	0.136	0.094	0.266
F_5	0.010	0.009	0.020
F_6	0.010	0.012	0.013
F_7	0.130	0.119	0.117
F_8	0.035	0.023	0.028
F_9	0.080	0.031	0.008
F_{10}	0.003	0.015	0.003
$Keijzer_1$	0.014	0.014	0.007
$Keijzer_2$	0.000	0.000	0.000
$Keijzer_3$	0.258	0.208	0.247
$Keijzer_4$	0.458	0.458	0.458
$Keijzer_5$	0.185	0.185	0.185
$Keijzer_6$	13.548	44.815	160.873

Table 4 Standard deviation of the results obtained for each instance and GE variant

Instance	Classical GE	Random permutations	Proposal
F_1	0.080	0.053	0.033
F_2	0.152	0.063	0.068
F_3	0.189	0.125	0.097
F_4	0.330	0.287	0.244
F_5	0.029	0.024	0.020
F_6	0.033	0.025	0.027
F_7	0.141	0.138	0.037
F_8	0.051	0.022	0.033
F_9	0.048	0.042	0.031
F_{10}	0.031	0.024	0.012
$Keijzer_1$	0.005	0.004	0.004
$Keijzer_2$	0.063	0.026	0.053
$Keijzer_3$	0.040	0.040	0.032
$Keijzer_4$	0.016	0.011	0.001
$Keijzer_5$	0.002	0.004	0.004
$Keijzer_6$	42.097	107.635	178.732

Table 5 Ranking based on Standard deviation

Algorithm	Ranking
Proposal	1.50
Random permutations	1.78
Classical GE	2.72

7 Conclusions and Future Work

Using grammatical evolution for symbolic regression can be a powerful approach to discovering mathematical equations that fit data, especially in situations where the underlying relationships are not well understood.

Although the non-parametric test over the median doesn't allow to decide which algorithm has the best performance, the non-parametric test with the standard deviation let to know that one of the GE variant is more consistent.

The proposal is more consistent and it's possible to say that applying the 2-opt to the index permutation is better than the random permutations or the classical GE.

With these results is possible to look up another kind of heuristics to mutate or make a local search based on permutations to obtain a GE more consistent and with better results.

Acknowledgements This work was partially supported by the National Council of Humanities, Science and Technology (CONAHCYT) of Mexico, via Postgraduate Scholarship 827730 (J. M. Perez-Rocha) and Grant CÁTEDRAS-2598 (A. Rojas).

References

1. G. Hintze, *Automatic Programming* (Springer, Berlin, Heidelberg, 1966), pp. 177–211. https://doi.org/10.1007/978-3-662-40151-4_7
2. R. Balzer, A 15 year perspective on automatic programming. IEEE Trans. Softw. Eng. **11**, 1257–1268 (1985)
3. U. Schmid, *6. Automatic Programming* (Springer, Berlin, Heidelberg, 2003), pp. 99–166. https://doi.org/10.1007/978-3-540-44846-4_6
4. J.R. Koza, Genetic Programming. On the Programming of Computers by Means of Natural Selection. Massachusetts Institute of Technology (1992)
5. W.B. Langdon, Genetic programming and data structures: genetic programming+ data structures= automatic programming! (1998). https://doi.org/10.1007/978-1-4615-5731-9
6. C. Ryan, J.J. Collins, M. O'Neill, Grammatical evolution—evolving programs for an arbitrary language. EuroGP (1998)
7. M. O'Neil, C. Ryan, *Grammatical Evolution* (Springer, Boston, MA, 2003), pp. 33–47. https://doi.org/10.1007/978-1-4615-0447-4_4
8. C. Ryan, M. O'Neill, J.J. Collins, *Handbook of Grammatical Evolution* (Springer, Cham, 2018), pp. 1–497. https://doi.org/10.1007/978-3-319-78717-6
9. M. O'Neill, C. Ryan, Grammatical evolution. IEEE Trans. Evol. Comput. **5**(4), 349–358 (2001). https://doi.org/10.1109/4235.942529

10. J.H. Holland, *Adaptation in Natural and Artificial Systems. An Introductory Analysis with Applications to Biology, Control, and Artificial Intelligence* (MIT Press, 1992)
11. M. O'Neill, A. Brabazon, et al., Grammatical differential evolution, in: IC-AI (2006), pp. 231–236
12. M. O'Neill, A. Brabazon, Grammatical swarm, in *Genetic and Evolutionary Computation–GECCO 2004: Genetic and Evolutionary Computation Conference* (Seattle, WA, USA, June 26-30, 2004. Proceedings, Part I, 2004), pp. 163–174
13. M.A. Sotelo-Figueroa, A. Hernández-Aguirre, A. Espinal, J.A. Soria-Alcaraz, J. Ortiz-López, Symbolic Regression by Means of Grammatical Evolution with Estimation Distribution Algorithms as Search Engine, in *Recent Advances in Harmony Search Algorithm*, vol. 749 (Springer, Cham, 2018), pp. 169–177. https://doi.org/10.1007/978-3-319-71008-2_14
14. B.V. Zúñiga, J.M. Carpio, M.A. Sotelo-Figueroa, A. Espinal, O.J. Purata-Sifuentes, M. Ornelas, J.A. Soria-Alcaraz, A. Rojas, Exploring random permutations effects on the mapping process for grammatical evolution. J. Autom., Mobile Robot. Intell. Syst. 65–72 (2020). https://doi.org/10.14313/JAMRIS/1-2020/8
15. D.A. Augusto, H.J.C. Barbosa, Symbolic regression via genetic programming. In: Proceedings of the VI Brazilian Symposium on Neural Networks (SBRN'00). SBRN '00. IEEE Computer Society, USA (2000), p. 173
16. P. Barmpalexis, K. Kachrimanis, A. Tsakonas, E. Georgarakis, Symbolic regression via genetic programming in the optimization of a controlled release pharmaceutical formulation. Chemometrics Intell. Laboratory Syst. **107**(1), 75–82 (2011). https://doi.org/10.1016/j.chemolab.2011.01.012
17. M.A. Sotelo-Figueroa, H.J. Puga Soberanes, J.M. Carpio Valadez, H.J. Fraire Huacuja, L. Cruz-Reyes, J.A. Soria-Alcaraz, Improving the Bin Packing Heuristic through grammatical evolution based on swarm intelligence. Math. Probl. Eng. **2014**(1), 1–12 (2014). https://doi.org/10.1155/2014/545191
18. I.M. Oliver, D.J. Smith, J.R.C. Holland, A study of permutation crossover operators on the traveling salesman problem, in *Proceedings of the Second International Conference on Genetic Algorithms on Genetic Algorithms and Their Application* (L. Erlbaum Associates Inc., USA, 1987), pp. 224–230
19. J.W. Backus, F.L. Bauer, J. Green, C. Katz, J. McCarthy, P. Naur, A.J. Perlis, H. Rutishauser, K. Samelson, B. Vauquois, J.H. Wegstein, A. Wijngaarden, M. Woodger, Revised report on the algorithmic language ALGOL 60. The Comput. J. **5**(4), 349–367 (1963). https://doi.org/10.1093/comjnl/5.4.349
20. B.V. Zuñiga-Nuñez, J.M. Carpio, M.A. Sotelo-Figueroa, J.A. Soria-Alcaraz, O.J. Purata-Sifuentes, M. Ornelas, A. Rojas-Domínguez, Studying grammatical evolution's mapping processes for symbolic regression problems, in *Intuitionistic and Type-2 Fuzzy Logic Enhancements in Neural and Optimization Algorithms: Theory and Applications* (Springer, 2020), pp. 445–459. https://doi.org/10.1007/978-3-030-35445-9_32
21. D. Karaboga, C. Ozturk, N. Karaboga, B. Gorkemli, Artificial bee colony programming for symbolic regression. Inform. Sci. **209**, 1–15 (2012). https://doi.org/10.1016/j.ins.2012.05.002
22. G.A. Croes, A method for solving traveling-salesman problems. Oper. Res. **6**(6), 791–812 (1958). https://doi.org/10.1287/opre.6.6.791
23. G. Laporte, M. Gendreau, J.-Y. Potvin, F. Semet, Classical and modern heuristics for the vehicle routing problem. Int. Trans. Oper. Res. **7**(4), 285–300 (2000). https://doi.org/10.1016/S0969-6016(00)00003-4
24. M. Keijzer, *Improving Symbolic Regression with Interval Arithmetic and Linear Scaling. In: Genetic Programming* (Springer, 2003), pp. 70–82. https://doi.org/10.1007/3-540-36599-0_7
25. D.R. White, J. McDermott, M. Castelli, L. Manzoni, B.W. Goldman, G. Kronberger, W. Jaśkowski, U.-M. O'Reilly, S. Luke, Better GP benchmarks: Community survey results and proposals. Genetic Programming and Evolvable Machines **14**(1), 3–29 (2013). https://doi.org/10.1007/s10710-012-9177-2
26. K.K. Gupt, M.A. Raja, A. Murphy, A. Youssef, C. Ryan, GELAB—the cutting edge of grammatical evolution. IEEE Access 1–1 (2022). https://doi.org/10.1109/ACCESS.2022.3166115

Evolutionary Deployment of Central Pattern Generators for Legged Robots Using Nengo

Ricardo Pérez-López, Andrés Espinal, Marco Sotelo-Figueroa, Erick I. Guerra-Hernandez, Patricia Batres-Mendoza, and Horacio Rostro-Gonzalez

1 Introduction

Since its inception, the field of robotics has been heavily influenced by nature. As a result, we now have robotic systems that can accurately mimic the behavior of various biological creatures. Locomotion is one of the main topics of interest within bio-inspired robotics, as it remains a partially solved engineering problem. Legged robotic systems, as those based on insect morphology, require a complex process of synchronizing many variables in parallel; in Fig. 1 there are shown three different legged robots. For instance, a hexapod robot typically has three degrees of freedom per limb, consisting of the coxa, femur, and tibia; which means that eighteen mechanisms must be controlled per time step for achieving its locomotion.

Although there are several techniques to address locomotion in legged robots, a remarkable approach is biological inspired through specialized neural circuits referred as Central Pattern Generators (CPGs); which are capable of producing rhythmic and stereotyped patterns without requiring input stimuli. In living beings,

R. Pérez-López
Departamento de Electronica, Universidad de Guanajuato, Carr. Salamanca-Valle de Santiago Km 3.5, 36787 Salamanca, Gto., Mexico

A. Espinal (✉) · M. Sotelo-Figueroa
Departamento de Estudios Organizacionales, Universidad de Guanajuato, Fracc. 1, 36250 Guanajuato, Gto., Mexico
e-mail: aespinal@ugto.mx

E. I. Guerra-Hernandez · P. Batres-Mendoza
Facultad de Sistemas Biologicos e Innovación, Universdad Autonoma Benito Juarez de Oaxaca, Av. Universidad S/N, 68120 Oaxaca, Oaxaca, Mexico

H. Rostro-Gonzalez
Grup d'Enginyeria en Producte Industrial (GEPI), Institut Quimic de Sarria, Universitat Ramon Llull, Via Augusta 390, 08017 Barcelona, Spain
e-mail: horacio.rostro@iqs.url.edu

P. Melin and O. Castillo (eds.), *New Directions on Hybrid Intelligent Systems Based on Neural Networks, Fuzzy Logic, and Optimization Algorithms*, Studies in Computational Intelligence 1146, https://doi.org/10.1007/978-3-031-53713-4_12

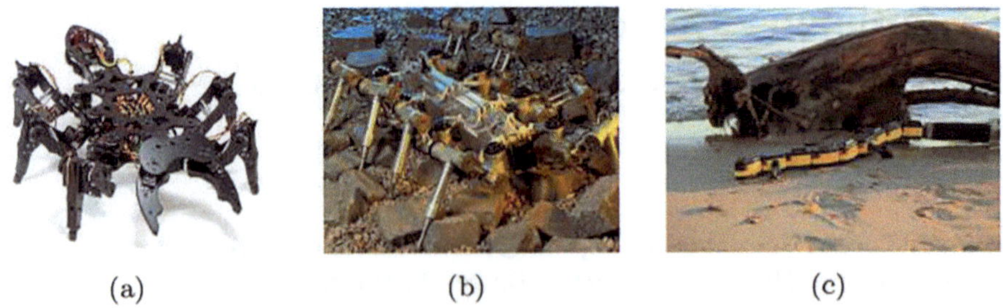

Fig. 1 **a** Ant robot [Image retrieved from the official site of the company Lynxmotion], **b** Scorpion robot [1], **c** Salamander robot [2]

as described in [3], the flexion and tension of the muscles involved in the walking action are promoted by neural mechanisms in which neurons inhibit each other. Furthermore, CPGs directly contribute to the unconscious movements that occur in vertebrates, including digestion, locomotion (walking, running, flying, etc.), and breathing [4]. From a mathematical perspective, CPG-based locomotion systems have been modeled with varying degrees of biological plausibility. A particular kind of CPGs have been defined by using Spiking Neural Networks (SNNs), these kind of networks are build by models (spiking neurons) that simulate the electrochemical processes taking place in the brain using processing units described by differential equations; the range of spiking neuron models goes from simplified ones incorporating integrate-and-fire neurons to very detailed models such as the Hodgkin-Huxley model [5].

Although there are alternative methodologies exist for designing CPGs, like coupled oscillators [6] that even though they provide good performance, they often require more resources and lack of biological plausibility. In contrast, the employment of SNNs in CPG design has gained momentum in recent years due to this kind of networks is more plausible and the emerging advancements in neuromorphic engineering. Neuromorphic boards have the ability to emulate the electrophysiological behavior of biological neurons, as well as real conductances. They are available through a hybrid architecture of digital and analog circuits; thus, this technology represents one of the most promising prospects for the hardware implementation of artificial neural networks.

In this work, we establish an evolutionary methodology to design CPGs based on SNNs. The design proposal is based on divide-and-conquer paradigm in which the evolutionary algorithm defines both synaptic connections and weights of a single spiking neuron in order to replicate an specific rhythmic and periodical signal of a locomotion pattern. This process is carried out for each different signal of a locomotion pattern, aftwards all individual designs are integrated into a SNN which in turns generates the whole locomotion pattern. By means of this methodology, there were generated CPGs for different locomotion pattern of quadruped and hexapod robots, achieving SNN with smaller architectures that similar reported in the state of the art. The CPGs were designed and implemented using Nengo simulator with

Loihi backend, which enables mapping of the neural design built in Python to a low-level language, simplifying the future implementation of artificial neural networks in hardware, specifically in the Intel's Loihi neuromorphic card.

The paper is structured as follows: Sect. 2 details the elements and materials used in the project's development. Section 3 outlines the methodology employed for the proposed design. Section 4 displays the attained results, and finally, Sect. 5 provides conclusions and future work.

2 Background

This section describes the models and components used to develop the proposal. First, the locomotion patterns of different gaits for quadruped and hexapod robots are described in Sect. 2.1. Next, in Sect. 2.2, the operation and characteristics of SNN-based CPG are detailed. Finally, the Evolutionary Algorithm is presented in Sect. 2.3.

2.1 Locomotion Patterns

The rhythmic patterns used to simulate locomotion in quadruped and hexapod robots are taken from [7], which are initially inspired on a biological study, where the coordination between the legs of the stick insect (carausius morosus) is described, as well as the gait types, which vary from a tripod coordination in fast walkers to a metachronous or undulating gait in slow walkers [8]. For practical purposes, three gait patterns for quadruped robots proposed in [9] are considered in this work, *Running* (Fig. 2a), *Jogging* (Fig. 2b) and *Walking* (Fig. 2c), and the same three gait patterns but for hexapod robots proposed in [7] are also used for this work, *Running* (Fig. 3a), *Jogging* (Fig. 3b) and *Walking* (Fig. 3c).

In Figs. 2 and 3, the x-axis represents the time scale for reference, while the y-axis corresponds to the robot joints (coxa and femur, as shown in Fig. 4a). The blue bars represent the neural activity stimulating the servomotors, while the blanks indicate their absence. Thus, the Coxa-Left (**CL**) and Coxa-Right (**CR**), numbered from 1 to 2 for the quadruped (Fig. 4a) and from 1 to 3 for the hexapod (Fig. 4b), respectively correspond to the servomotors for the coxa, which is responsible for the movement of the hip joint from back to front and vice versa. On the other hand, the servomotors of the femur, which are responsible for the movement from bottom to top and vice versa, are represented by Femur-Left (**FL**) and Femur-Right (**FR**) ranging from 1 to 2 for the quadruped and from 1 to 3 for the hexapod.

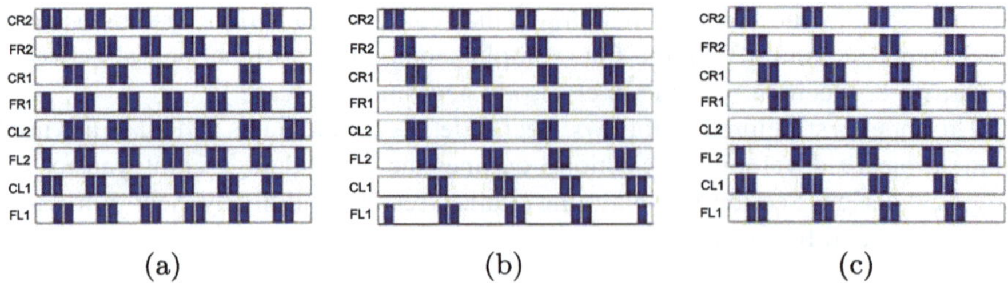

Fig. 2 Locomotion patterns for a quadruped robot proposed in [7] (taken from [10]): **a** Running, **b** Jogging and **c** Walking

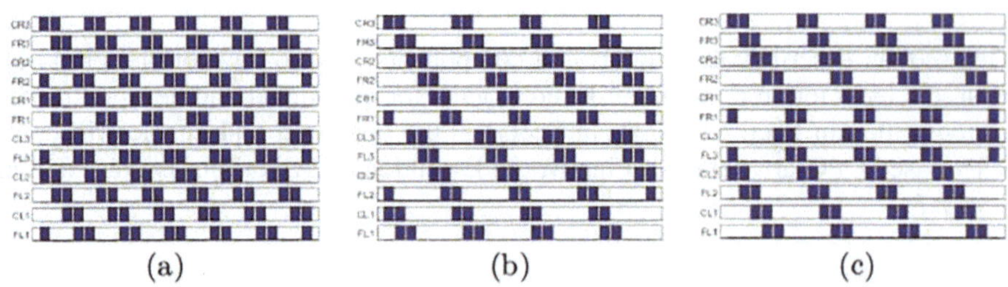

Fig. 3 Locomotion patterns for a hexapod robot proposed in [7] (taken from [10]): **a** Running, **b** Jogging and **c** Walking

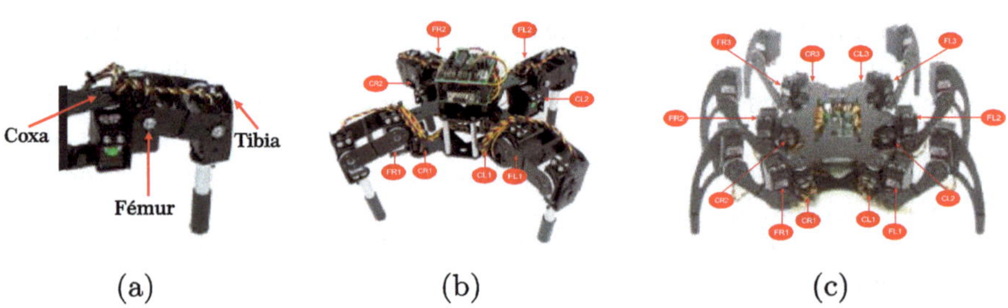

Fig. 4 **a** Extremity with three degrees of freedom **b** Quadruped robot and **c** Hexapod robot

2.2 Spiking Central Pattern Generator

In vertebrates, the central nervous system consists of neural circuits that communicate with each other to respond to external and internal stimuli and generate output signals to meet the organism's needs. Locomotion, as well as other involuntary activities, require the generation of rhythmic and stereotyped signals to stimulate various muscle groups; these signals are known as motor patterns. The patterns originated for unconscious movements are created by specialized neural circuits called Central Pattern Generators (CPGs) [11, 12].

A CPG is a neural network that endogenously generates rhythmic signals without intervening in external sensory information. These systems are responsible for exe-

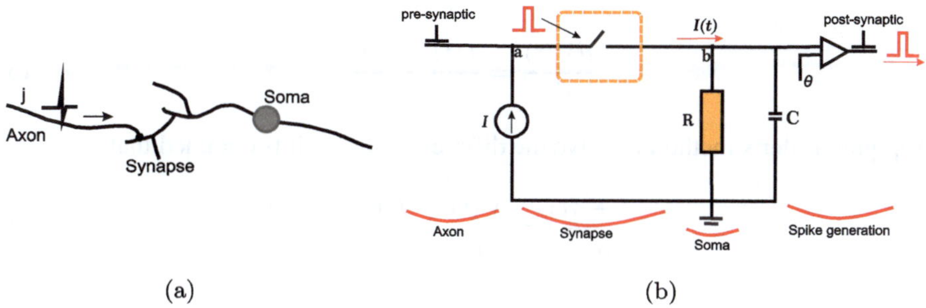

Fig. 5 Schematic diagram of the integrate-and-fire neural model

cuting various unconscious activities in living organisms, such as breathing, digestion, heartbeat, locomotion, etc. In robotics, these neural systems have already been successfully implemented as SNNs for gait control in legged robots (e.g., hexapods, quadrupeds, bipeds, etc.) [10].

SNNs are plausible models that can accurately emulate the realistic behavior of biological neurons. In a spiking neuron, the output is not only determined by the input (spatial dependence) but also by time. This allows the neuron to change its state even without input stimuli and to incorporate spatiotemporal information through neuronal connectivity and plasticity, just like real neurons [13]. The amplitude of input or output spikes is not informative. Instead, the temporal moments of spikes contain all the necessary information.

Different mathematical models have been proposed to describe their neuronal activity, such as the Leaky Integrate-and-Fire (LIF) model [14]. This model can replicate the neuronal dynamics observed in the brain through an Resistor-Capacitor (RC) circuit and describe the state of a neuron about its membrane potential, which is determined by the synaptic inputs and injected currents from the neurons it receives (Fig. 5a).

In the circuit of Fig. 5b, the current $I(t)$ charges the RC circuit. Subsequently, the voltage across the capacitor is compared with the threshold voltage θ. If the threshold is exceeded at a given time, an output pulse is generated. At node b, the current $I(t)$ is divided into two components: $I(t) = I_R + I_C$ where I_R represents the current in the resistor R and can be calculated by Ohm's Law as $I_R = V(t)/R$. While the second component I_C charges the capacitor C and is defined as $I_C = CdV/dt$, Therefore:

$$I(t) = \frac{V(t)}{R} + C\frac{dV(t)}{dt} \tag{1}$$

clearing

$$\frac{dV(t)}{dt} = \frac{I(t)}{C} - \frac{V(t)}{RC} \tag{2}$$

replacing $\tau = RC$

$$\frac{dV(t)}{dt} = \frac{I(t)}{C} - \frac{V(t)}{\tau} \tag{3}$$

Applying Euler's method to solve the differential Eq. 3, it is obtained that

$$\frac{V(t+dt)}{dt} = \frac{V(t)}{dt} + \frac{I(t)}{C} - \frac{V(t)}{\tau} \tag{4}$$

2.3 Evolutionary Strategy

Evolutionary algorithms are techniques based on the Darwinian paradigm of evolution, whose fundamental principles are variation and selection. This concept, coupled with generational change, constitutes the essential components of the evolutionary cycle [15]. Darwinian evolution is often considered similar to the optimization concept, so algorithms inspired by it are predominantly applied to optimize fitness functions, such as the evolutionary strategy (ES) algorithm.

The Evolutionary Strategy Algorithm (ES) is an optimization technique that draws inspiration from biological evolution and finds extensive use in artificial intelligence. The approach works on the principles of natural selection, mutation, and recombination to enhance candidate solutions in optimization problems. In ES, the candidate solutions population evolves over several generations and tends to converge towards high-quality solutions in the search space [16]. There are various Python frameworks available for implementing optimization algorithms to solve fitness functions. For this study, the Pymoo library's ES method was utilized [17].

3 Methodology

This section describes the implementation of the SNN-based CPG in the Nengo simulator and the parameter estimation by means Evolutionary Strategy algorithm.

3.1 Parameter Estimation

Parameter estimation (synaptic connections and weights) of a SNN is not a simple task. In this work, there are filtered the different signals of each locomotion gait (described in Sect. 2.1) which allows to generate compact architectures avoiding repeated signals. The different signals of a locomotion pattern are injected individually to spiking neurons as binary vectors [0, 1]. This information is then processed following the divide-and-conquer strategy, which states that a complex problem can be divided into subproblems that are solved independently and then combined into an ensemble solution [18]. That is, the SNN architecture is divided into individual connectivity designs rather than building and training an SNN as a whole. Here, each neural connectivity design is implemented using an evolutionary method that searches for an adequate configuration for a specific neuron. In this way, the solution is produced when each neural subsystem, based on an evolutionary approach, encodes the information into solutions of connections and synaptic weights, ensuring that each of them periodically replicates a target rhythmic pattern for robot walking. The final CPG is obtained by integrating all the individual neuron designs.

3.1.1 Input Coding

The inputs are constructed using the locomotion patterns shown in Figs. 2 and 3. These patterns are coded as binary logic sequences, where each bar (neuronal activation) is represented by a logical "1" and the empty spaces (absence of neuronal activity) by a logical "0". Figure 6 illustrates the encoding of the input patterns for the *Running* gait of a quadruped robot. This shows that some of the sequences are repeated, such as **CR2** = **CL1**, **FR2** = **FL1**, **CR1** = **CL2** and **FR1** = **FL2**, allowing to reduce the number of spiking neurons needed in the process of parameter estimation and subsequent CPG integration. That is, instead of having eight spiking neurons corresponding to the number of input logic sequences to be replicated, it is sufficient to have four of them to generate the eight signals, which translates into a saving of resources. This approach was also applied to the *Jogging* and *Walking* gaits, reducing the number of spiking neurons from eight to six. In the case of the hexapod robot, a more favorable optimization was attained regarding the number of required neurons, as it was reduced from 12 to 4 for *Running* gait and from 12 to 6 for the *Jogging* and *Walking* gaits. Therefore, it is sufficient to replicate one of the duplicated sequences to emulate the behavior of two or more joints, enabling a single neuron to control the movement of two or three servomotors concurrently in the robots.

To define an input in Nengo, it is necessary to use the *Node* object, which functions as a type of coupler that allows connecting non-neural inputs (signals, functions,

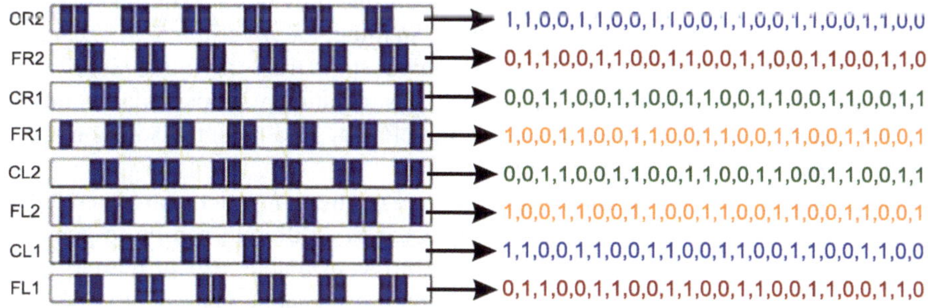

Fig. 6 Coding of the *Running* gait for a quadruped

Table 1 Basic parameters of a neuron in Nengo

Parameters	Value	Comments
Neurons	1	# No. neurons (LIF)
Dimensions	1	# No. dimensions
Intercepts	[−0.0001]	# Neurons activation
Max_rate	[400]	# Maximum firing rate (Hz)
Encoders	[1]	# Encoders
Neuron_type	Nengo . LIF()	# Neural model

numerical arrays, etc.) to Nengo objects. *Nodes* do not part of a brain model but serve to summarize assumptions made about sensory data or other environmental variables. With this in mind, each binary vector is input to a *Node* to be recognized as input to the Nengo neural models. In addition, the reading period of the input data is specified by the instruction *nengo.processes.PresentInput*, which, in this case, was 0.002 ms.

Once the binary vectors have been transformed into Nengo inputs, the internal parameters governing the behavior of a LIF spiking neuron in Nengo are set such that the neuron is able to generate spikes with the same temporal characteristics as the injected inputs (Table 1).

3.1.2 Method for Parameter Estimation

After determining the parameters for defining a spiking neuron in Nengo, we create the neural model architecture to estimate SNN parameters. The configuration of each neuron is carried out by the metaheuristic algorithm, called Evolutionary Strategy, which uses a fitness function in order to explore within a search space composed of all weighted connectivity configurations. Such fitness function is same as used in [9], the *Bivariate Spike-Distance* [19] is a temporal average of the dissimilarity profiles between the spike trains, serves as a unit of measurement for estimating the degree of synchrony between two spike trains. A value of 0 denotes total synchrony between the analyzed spike trains, while any other recorded value provides a quantified measure of dissimilarity ranging from 0 to 1. With this in mind, the *Bivariate Spike-Distance* is used to measure the similarity between a target spike train (rhythmic signal to be replicated) and the spike train generated after the simulation of the candidate solution. In this manner, it is feasible to determine the efficacy of a potential solution; i.e. a spiking neuron linked to its presynaptic neurons.

The design process is described as follows: Locomotion patterns are encoded as binary sequences and injected individually, each to a different neuron to generate spike trains. Then, these spike trains are set up as available, except the target, to generate presynaptic neuronal activity with the postsynaptic neuron. It is decided whether, at time $t = 0$, the neuron should fire or not. Subsequently, the evolutionary

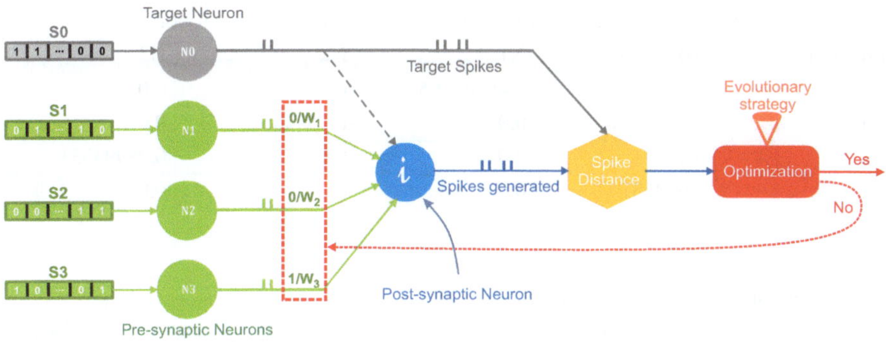

Fig. 7 Diagram of the parameter estimation method for the *running* gait

strategy generates candidate solutions for connectivity and synaptic weights until the expected error of 0 is reached. This same process is repeated continuously until each of the input spike trains is replicated. Finally, all the neural connectivity designs are stored to build the resulting CPG topology later.

Figure 7 shows the diagram of the design method described above for estimating the connections ($\mathbf{C}_1$, $\mathbf{C}_2$, and $\mathbf{C}_3$) and synaptic weights ($\mathbf{W}_1$, $\mathbf{W}_2$, and $\mathbf{W}_3$) of the SNN. Here, each binary input (S0, S1, S2, and S3) is injected to a presynaptic neuron (N0, N1, N2, and N3), three of which (green) have neuronal activity available. Then, the presynaptic neurons N1, N2, and N3 are subjected to a conditional rule where the level of importance of each neuron to replicate the pulse pattern generated by the target neuron is measured. In that case, the synaptic connection between the current presynaptic neuron and the postsynaptic neuron *Ng* (blue) is considered necessary ($\mathbf{C}_n = 1$); otherwise, no connection is made, remaining exclusive in the system ($\mathbf{C}_n = 0$). Subsequently, both the spike train produced by the *Ng* neuron and the one generated by the target neuron are compared using the Bivariate SPIKE-distance. The resulting value is then optimized using an evolutionary strategy method, which evolves candidate connectivity solutions until the expected error (*spike distance* = 0) is achieved. After replicating the pulse train of the target neuron, the resulting connectivity solution created by the contributing neurons is stored. This process is repeated with the remaining neurons until the *Running* gait's distinct locomotion patterns are successfully replicated.

4 Results

This section details the integration of each CPG of the different types of (*Running*, *Jogging*, and *Walking*) gaits for both quadruped and hexapod robots.

As an example, we explain the resultos for *Running* gait of quadruped robot, incluiding the of the weights matrix, topology, and simulation of the designed CPG.

Table 2 Neural connectivity matrix for the *running* gait of the quadruped robot

Pre\post	N0	N1	N2	N3
N0	0.0	0.48761765	0.0	0.0
N1	0.0	0.0	0.25945331	0.0
N2	0.0	0.0	0.0	0.98994331
N3	0.6996376	0.0	0.0	0.0

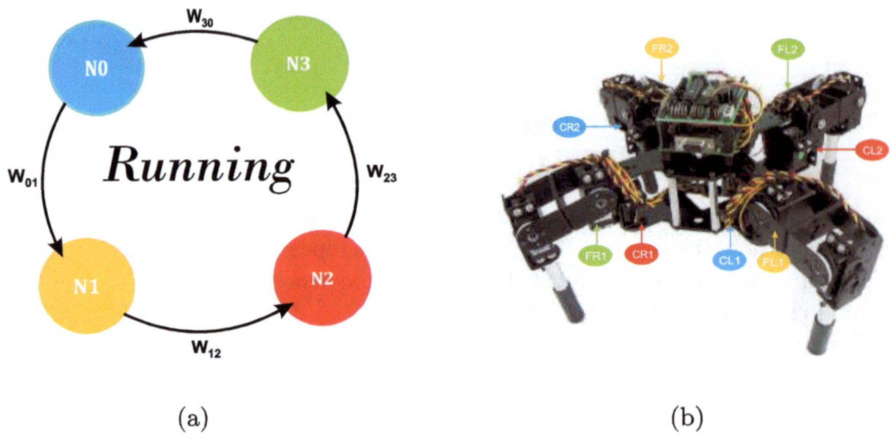

Fig. 8 **a** CPG topology for *running* gait, **b** Leg joints of the quadruped robot

Table 2 shows the connectivity matrix and synaptic weights resulting from the evolutionary process. Each row corresponds to a presynaptic neuron, and each column corresponds to a postsynaptic neuron. The presence of non-zero values indicates a weighted synaptic connection between a presynaptic neuron and a postsynaptic neuron, which can be either excitatory (positive) or inhibitory (negative). Moreover, as can be seen, only one synaptic connection is needed between each neuron, which considerably simplifies the topologies of the different CPGs.

The connectivity matrix in Table 2 yielded the topology of the CPG for the *Running* gait, as depicted in Fig. 8a; here, each neuron is indicated by a specific color and distinguishes the joints of the quadruped robot being controlled by that neuron, as presented in Fig. 8b. Synaptic connections are represented by black arrows, which originate from presynaptic neurons and terminate at postsynaptic neurons. In addition, the letter **W** represents the synaptic weights with two numerical subscripts. The first subscript represents the presynaptic neuron of the connection, while the second represents the postsynaptic neuron. For instance, using $\mathbf{W}_{30}$ would refer to the synaptic weight between the connection neuron 0 and neuron 3 ($\mathbf{W}_{30} = 0.48761765$). Ultimately, the CPG was determined by four neurons with excitatory synaptic connections.

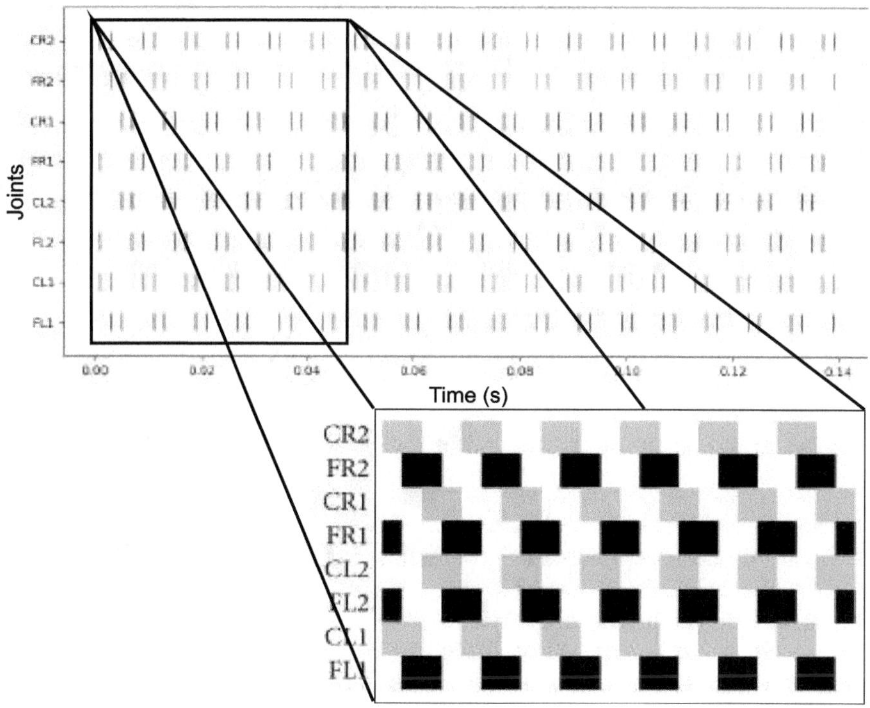

Fig. 9 CPG simulation for *jogging* gait of the quadruped robot

Figure 11 presents the Python simulation of the Central Pattern Generator (CPG) designed for the *running* gait of the quadruped robot. As can be seen, when the simulation reaches time 0.048 s, the locomotion pattern produced is identical to the pattern to be replicated. In addition, it can be seen that the CPG can continue generating the locomotion pattern indefinitely, giving validity to the system.

Similarly, results were obtained for the different gait types used in the quadruped and hexapod robots. To avoid redundancies, Figs. 9, 10, 12, 13, and 14 show the simulation of each designed CPG.

5 Conclusions and Future Work

In this work, we showed how to implement a CPG based on SNNs to control the gait of two bio-inspired robots, which can reproduce periodic locomotion patterns observed in some living organisms, such as the stick insect. In addition, the objective of replicating three different types of gait (*walking*, *jogging*, and *running*) was met through an evolutionary metaheuristic and using a minimum number of neurons in the designs. Likewise, as the architectures were designed using the Nengo simulator,

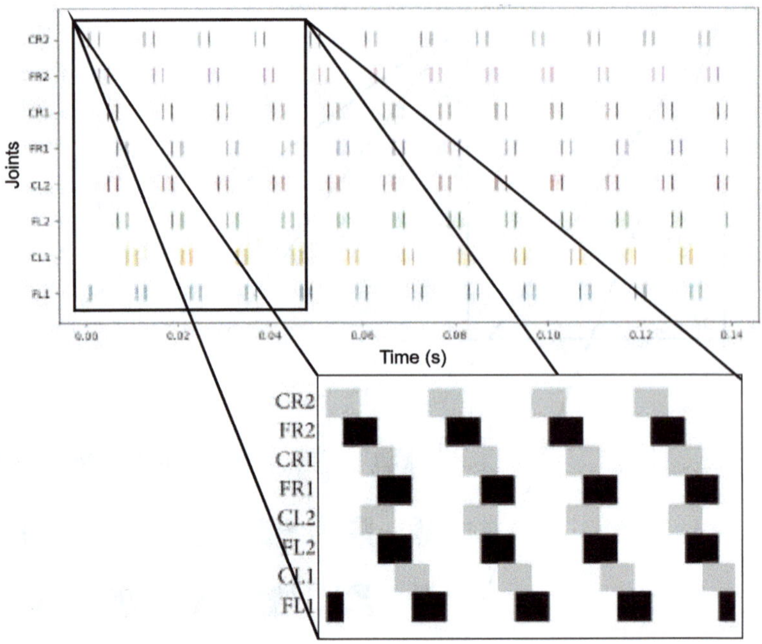

Fig. 10 CPG simulation for *walking* gait of the quadruped robot

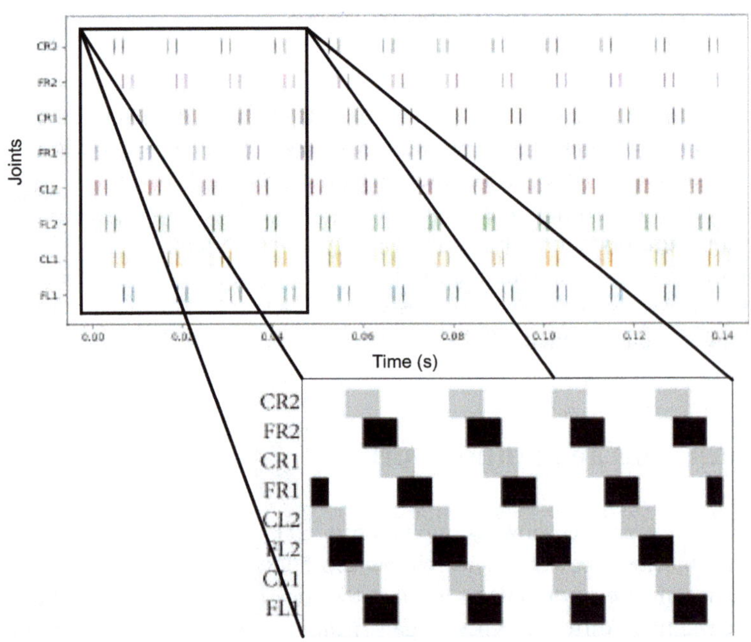

Fig. 11 CPG simulation for *running* gait of the quadruped robot

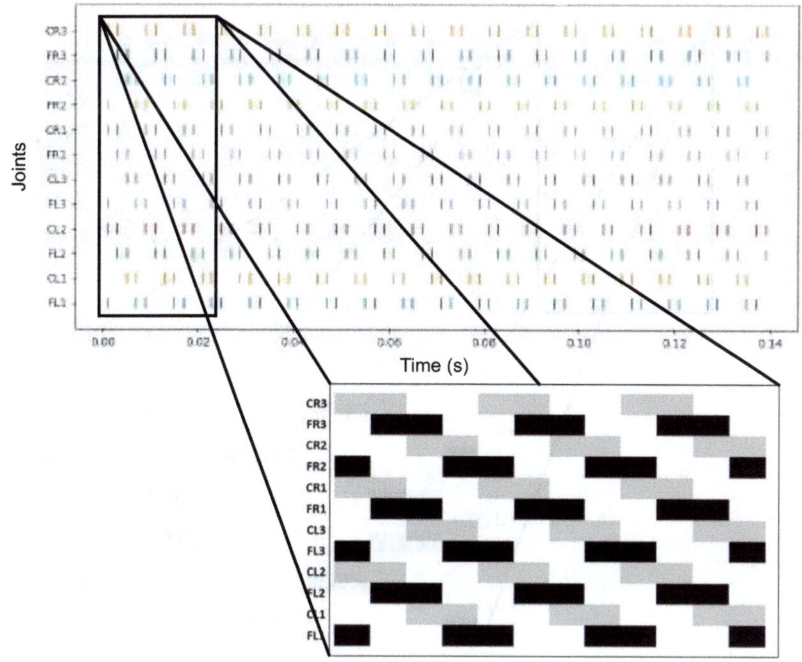

Fig. 12 CPG simulation for *running* gait of the hexapod robot

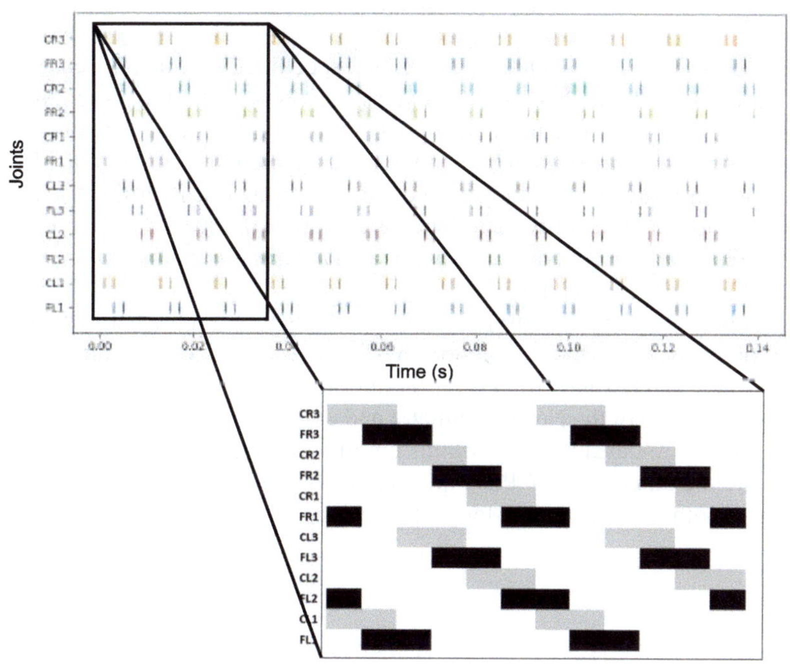

Fig. 13 CPG simulation for *jogging* gait of the hexapod robot

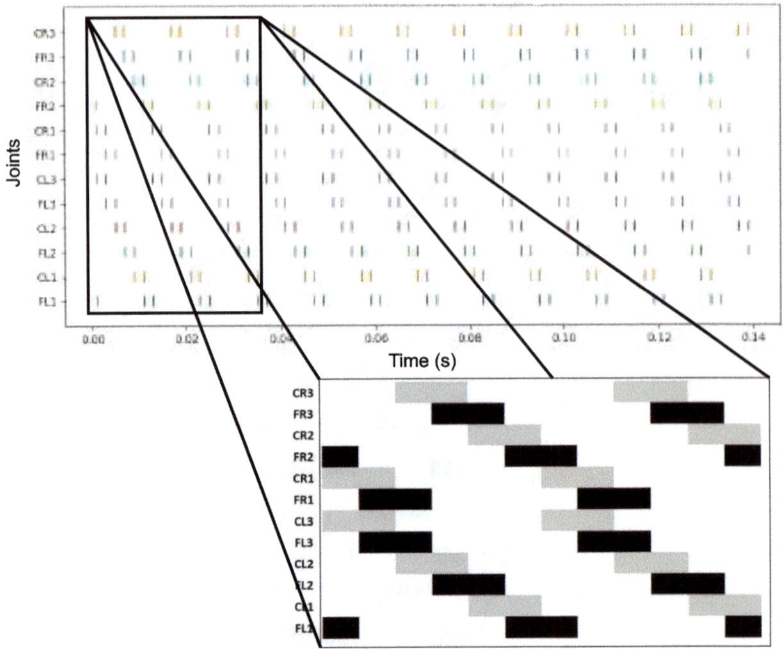

Fig. 14 CPG simulation for *walking* gait of the hexapod robot

a future hardware implementation is feasible, specifically on Intel's neuromorphic *Loihi* card. Thanks to the *Nengo_Loihi* extension, it is possible to map neural designs created in a high-level language, such as Python, to a low-level language without the direct need to transform and adapt the codes manually. Currently, this remains a challenge in artificial intelligence since, in most cases, it is difficult to land the algorithms designed in software to hardware. This is the reason for the interest of this research: to explore alternative ways to facilitate this process using computational tools such as the Nengo simulator.

As future work, implementing the CPG in the neuromorphic hardware *Loihi* of Intel is sought to control the locomotion physically (walking, jogging, and running) of the quadruped and hexapod robots. Furthermore, we aim to investigate the functionalities of *Loihi* tool of the Nengo simulator and expand the search for simulators that can map software-developed algorithms onto neuromorphic cards. Finally, we aim to compare the efficiency of CPG implementation on Loihi neuromorphic hardware with that on other embedded systems like FPGAs.

References

1. Robotics innovation center, German research center for artificial intelligence Gmbh. www.robotik.dfki-bremen.de/en/research/projects/aramies.html,2005-2007
2. A. Crespi, K. Karakasiliotis, A. Guignard, A.J. Ijspeert, Salamandra robotica ii: an amphibious robot to study salamander-like swimming and walking gaits. IEEE Trans. Robot. **29**, 308–320 (2013)
3. T.G. Brown, The intrinsic factors in the act of progression in the mammal. *Proceedings of the Royal Society of London. Series B, Containing Papers of a Biological Character* **84**, 308–319 (1911)
4. R. Brette et al., Simulation of networks of spiking neurons: a review of tools and strategies. J. Comput. Neurosci. **23**, 349–398 (2007)
5. A.J. Ijspeert, A. Crespi, D. Ryczko, J.-M. Cabelguen, From swimming to walking with a salamander robot driven by a spinal cord model. Science **315**, 1416–1420 (2007)
6. Z.G. Zhang, Y. Fukuoka, H. Kimura, Stable quadrupedal running based spring-loaded two-segment legged on a model **3**, 2601–2606 (2004)
7. H. Rostro-Gonzalez et al., A CPG system based on spiking neurons for hexapod robot locomotion. Neurocomputing **170**, 47–54 (2015)
8. M. Grabowska, E. Godlewska, J. Schmidt, S. Daun-Gruhn, Quadrupedal gaits in hexapod animals-inter-leg coordination in free-walking adult stick insects. J. Exp. Biol. **215**, 4255–4266 (2012)
9. A. Espinal, *et al.* Quadrupedal robot locomotion: a biologically inspired approach and its hardware implementation. *Computational Intelligence and Neuroscience* **2016** (2016)
10. E.I. Guerra-Hernandez et al., A fpga-based neuromorphic locomotion system for multi-legged robots. IEEE Access **5**, 8301–8312 (2017)
11. Duysens, H. W., Jacques y Van de Crommert. The central pattern generator from cats to humans. Neural control of locomotion; part 1. Gait & Posture **7**, 131–141 (1998)
12. T.G. Brown, On the nature of the fundamental activity of the nervous centres; together with an analysis of the conditioning of rhythmic activity in progression, and a theory of the evolution of function in the nervous system. The J. Physiol. **48**, 18 (1914)
13. Iakymchuk, T. *Spiking Neural Networks models targeted for implementation on Reconfigurable Hardware*. Ph.D. thesis, University Of Valencia (2017)
14. W. Gerstner, Time structure of the activity in neural network models. Phys. Rev. E **51**, 738 (1995)
15. H.-G. Beyer, *The Theory of Evolution Strategies* (Springer Science & Business Media) (2001)
16. I. Rechenberg, Evolution strategy: nature's way of optimization 106–126 (1989)
17. J. Blank, K. Deb, Pymoo: multi-objective optimization in python. IEEE Access **8**, 89497–89509 (2020)
18. A. Espinal et al., Design of spiking central pattern generators for multiple locomotion gaits in hexapod robots by Christiansen grammar evolution. Front. Neurorobot. **10**, 6 (2016)
19. T. Kreuz, D. Chicharro, C. Houghton, R.G. Andrzejak, F. Mormann, Monitoring spike train synchrony. J. Neurophysiol. **109**, 1457–1472 (2013)

Meta-Heuristics

A Review on Dolphin Swarm Algorithm: Applications in Computational Intelligence

Fevrier Valdez

1 Introduction

Collective intelligence has become a hot topic in artificial intelligence and swarm intelligence fields since its appearance until to date. Several algorithms have been developed inspired on collective and swarm intelligence. Currently, particle swarm optimization [1], genetic algorithm [2], bee colony optimization [3, 4] firefly algorithm, dragonfly algorithm [5], DSA [6, 7], Bat algorithm [8, 9], Ant colony optimization [10], to mention a few, are well implemented and have shown favorable results. This chapter relates the study about Dolphin Swarm Algorithm to observe the developed applications based on this algorithm for solving optimization problems. However, in Table 1 is presented in chronological order of the most popular bio-inspired algorithms in the history. From Scopus database we collected the records with DSA. In Fig. 1, is presented a preferred reporting items for systematic reviews and meta-analyses (PRISMA) [11] diagram to understand this query. The PRISMA diagram shows the systematic review of papers related to the DSA, an can be seen the selection and inclusion process for a better understanding of its application and development in computational intelligence; where 80 records in total were found with the query "Dolphin Swarm Algorithm": where only 78 were considered for this study and 2 were removed because are pre-print papers.

The main contribution of this paper, is to observe the existing applications using the Dolphin Swarm Algorithm. Although, not many authors have been using since was developed; with this review, its possible to analyze the main applications where this optimization method can be applied.

F. Valdez (✉)
Division of Graduate Studies, Tijuana Institute of Technology Tijuana/Tecnologico Nacional de Mexico, Calzada Tecnologico S/N, 22414 Tijuana, BC, México
e-mail: fevrier@tectijuana.mx

P. Melin and O. Castillo (eds.), *New Directions on Hybrid Intelligent Systems Based on Neural Networks, Fuzzy Logic, and Optimization Algorithms*, Studies in Computational Intelligence 1146, https://doi.org/10.1007/978-3-031-53713-4_13

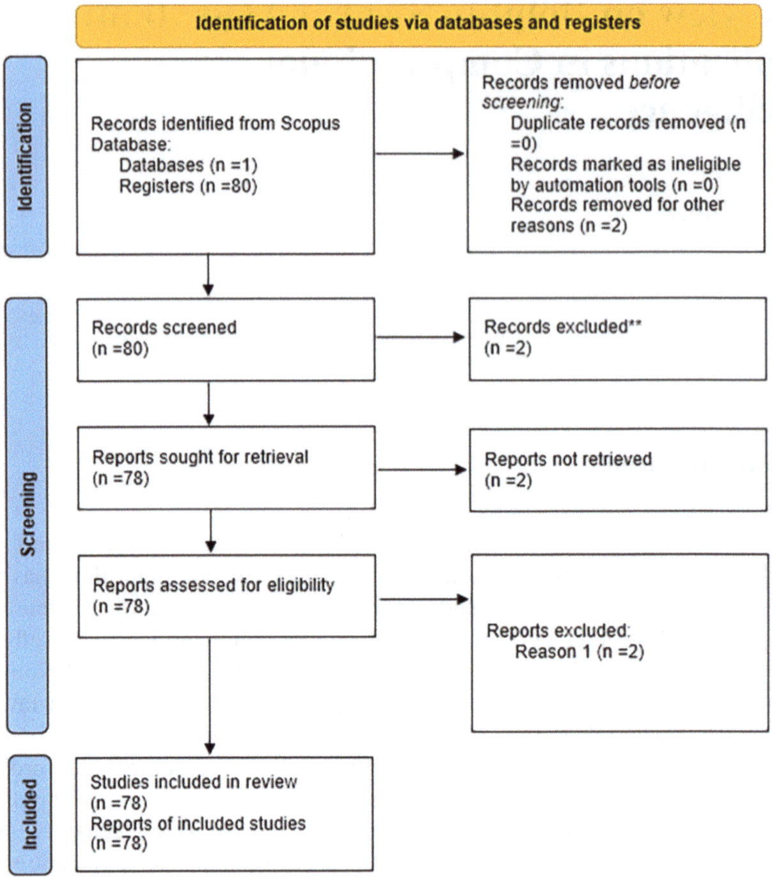

Fig. 1 PRISMA flow diagram

This paper is organized as follows: in Sect. 1 a brief introduction about of swarm intelligent is presented, Sect. 2 describes the literature review from Scopus database, Sect. 3 shows the DSA method analyzed in this study, in Sect. 4 are presented the applications in computational intelligence and finally the conclusions and future works are included in the last section.

2 Literature Review

In this Section, present the works with the DSA in the literature. From Scopus, we review the publications and authors that we are working with this optimization method. In [25] a fusion of dolphin swarm optimization and improved sine cosine algorithm for automatic detection and classification of objects from surveillance

Table 1 Popular bio-inspired algorithms

Year	Algorithm
1975	Genetic algorithms [12]
1992	Ant colony optimization [13]
1995	Particle swarm optimization [1]
1997	Tabu search [14]
2005	Artificia bee colony algorithm [15]
2006	Glow-worm [16]
2006	Cat swarm optimization [17]
2009	Cuckoo search [18]
2010	Firefly algorithm [19]
2010	Bat algorithm [20]
2016	Dolphin swarm algorithm [21]
2017	Optimization metaheuristic based on the self-defense of plants [22]
2020	Coronavirus optimization algorithm [23]
2022	Mycorrhiza optimization nature-inspired algorithm [24]

videos is presented. In this work, the authors proposed a dolphin swarm optimization (DSO) algorithm to detects and classify the objects from the surveillance videos. Also, in [26] an acoustic Detection and Recognition of Dolphins using Swarm Intelligence Neural Networks was proposed. On the other hand, in [27] DSO algorithm for software-defined antenna selection algorithm in underwater acoustic sensor network is presented. An application in scheduling it can be seen in [28]. Also, a recent work is presented in [29] different bio-inspired algorithms were used to reduce the network latency of underwater acoustic sensor. Also, in [28] and Application project task scheduling using dolphin swarm algorithm. By the other hand, in [30] a Dolphin echolocation approach for small-signal stability analysis of wind power system is presented. Also, in [31] is presented a software for cost estimation based on DSA. In [32] an Improved Dolphin Swarm Algorithm Based on Kernel Fuzzy C-Means is presented. Other improved DSA is presented in [33], where de authors proposed a modified DSA Based on Chaotic maps for solving high-dimensional function optimization problems. Also, in [34] an improved dolphin swarm optimization algorithm based on information entropy. Other important application in cryptoanalysis using DSA is presented in [35]. A DSA using a hybrid method is shown in [36]. Finally, other bio-inspired algorithms can be reviewed in [37].

3 Dolphin Swarm Algorithm

DSA, takes inspiration from the collective and coordinated behavior of dolphins in search and optimization tasks, which is based on mimicking the mechanism of dolphins in detecting, chasing after, and preying on swarms of sardines, squid, shrimp and others small marine animals. Dolphins exhibit highly organized group dynamics, echolocation, and collective cooperation, which have been translated into artificial methods. Figure 2 shows the dolphins collective behavior hunting squids.

The DSA has the following important concepts:

Echolocation: Dolphins use echolocation to locate prey and navigate in their environment. In the DSA, each dolphin (solution candidate) communicates and updates its position based on the information it receives from its peers.

Coordination: Dolphins often work together in groups to achieve common goals. Similarly, the DSA employs a collaborative approach where dolphins cooperate to find the optimal solution.

Adaptation: Dolphins adapt their strategies based on environmental changes. The DSA allows solutions to adapt through local search and optimization techniques.

The DSA has shown promise and versatility in addressing various optimization problems, including function optimization, feature selection, and parameter tuning. Its current state involves ongoing research and development aimed at improving its performance and applicability. Researchers have extended the DSA by incorporating additional features like dynamic parameter adaptation and hybridization with other algorithms.

Fig. 2 Dolphins collective behavior hunting squids

The future of the Dolphin Swarm Algorithm is bright. As it stands, this algorithm has demonstrated competence in various domains, but there is still considerable room for growth and innovation. Here are some directions for its future development:

Enhanced Adaptation: Improving the DSA's adaptability in dynamic environments is a crucial research area. Developing strategies that allow it to quickly adjust to changing problem landscapes will be valuable.

Hybrid Approaches: Combining the DSA with other bio-inspired or traditional optimization methods can potentially enhance its problem-solving capabilities.

Parallelization: Leveraging the power of parallel computing can help the DSA tackle larger and more complex optimization problems.

4 Applications in Computational Intelligence with DSA

The DSA has been used in several areas, such as:

Optimization mathematical functions: Solving complex mathematical functions efficiently.

Feature Selection: Identifying relevant features in large datasets, reducing dimensionality, and improving machine learning model performance.

Parameter Tuning: Optimizing hyperparameters for machine learning algorithms.

Data Clustering: Grouping similar data points for better understanding and analysis.

Routing and Network Optimization: Efficiently routing data in communication networks.

In this Section is presented the queries that we made from Scopus database. After reviewing the collected records from PRISMA diagram. The collected and filtered records, we used the Vosviewer [38] software, to make the corresponding networks and relations with authors, keywords, countries and other relevant information. Figure 3, shows the networks with the query 'Dolphin Swarm Algorithm'. Figure 4, shows the network density corresponding to the Fig. 3, with the aim of represent in the best way the generated network. In this case, we type the keywords in the field search from Scopus. As a result, 80 records were obtained. However, for this study two records were discarded, because they are in pre-print status. Subsequently, the 78 records were exported in a CSV file to build the network in Vosviewer. The generated network shows in Fig. 3 was built under the following characteristics:

Type of analysis: Co-occurrence.

Unit of analysis: All keywords.

Counting method: Full counting.

Number of the keywords selected: 138.

Minimum number of occurrences of a keyword: 2. Of the 874 keywords, 138 meet the threshold. For each 138 keywords, the total strength of the co-occurrence links with other keywords were calculated. The keywords with the greatest total link strength will be selected. Table 2, shows the Top 15 keywords with the greatest link.

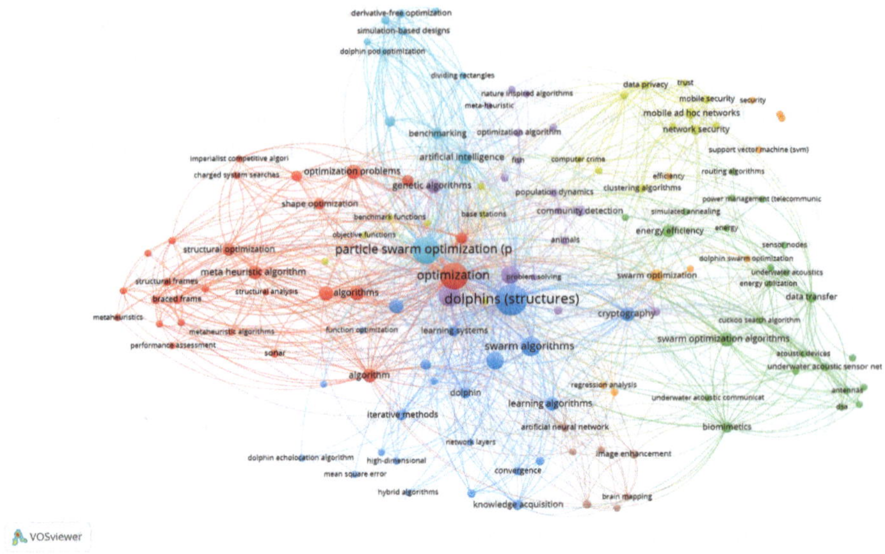

Fig. 3 Query from Scopus for the keywords 'dolphin swarm algorithm'

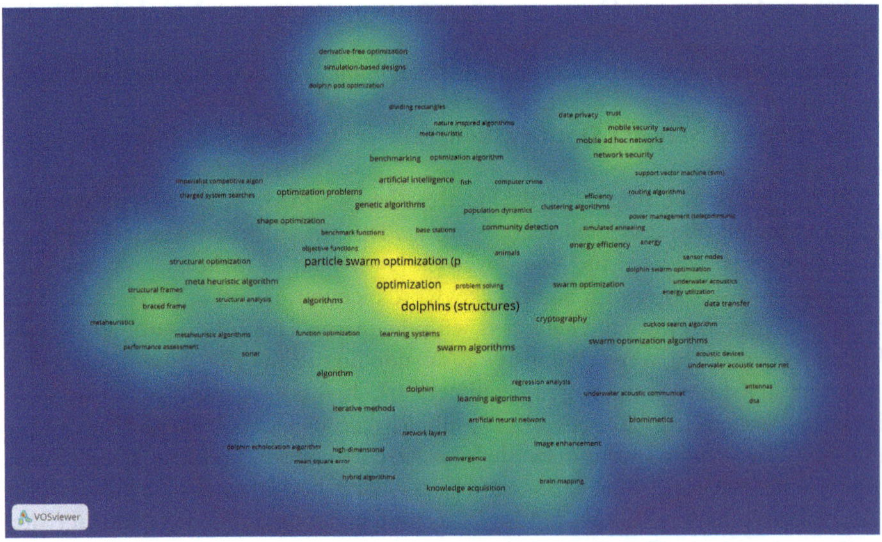

Fig. 4 Network density for the keywords 'dolphin swarm algorithm'

In this Table, it can be seen that the word with the most occurrences and strength is 'Dolphin (structure)'.

Table 2 Top 15 keywords with the greatest strength

Keyword	Occurrences	Total link strength
Dolphin (structure)	37	319
Optimization	28	262
Particle swarm optimization	28	225
Optimization algorithms	17	136
Swarm algorithms	14	103
Swarm intelligence	13	93
Dolphin swarm algorithm	12	90
Genetic algorithms	8	85
Evolutionary algorithms	8	84
Meta heuristic algorithm	7	80
Cryptography	8	79
Artificial intelligence	7	75
Algorithm	7	74
Learning algorithms	7	72
Swarm optimization algorithms	8	72
Optimization problems	8	71

5 Conclusions and Future Works

After reviewing the Dolphin Swarm Algorithm and its applications from Scopus, we can conclude that in the last years it can be noted that the use of the method is more frequent, many researchers currently base their work inspired by this method. It can also be highlighted that others choose to combine the method with other similar ones, achieving high performance with respect to the individual method. On the other hand, the main disadvantage to use this type of optimization methods, is the parameter setting, because the authors run the algorithms based with default values. However, others using parameter adaptation for solving this problem and the obtained results can be improved considerably. This paper help us to observe, the trends for using DSA in different applications. However, not many authors have attempted to use this optimization method; maybe because other techniques are more popular than DSA. As a future work, this review can be extended reviewing other bio-inspired optimization methods as Firefly, Dragonfly and others, like [39–42]. Also, other type of queries can be made with other database as Web of Science; for example, by authors, keywords, occurrences, etc.

Acknowledgements The author would like to thank CONAHCYT and Tecnológico Nacional de Mexico/Tijuana Institute of Technology for the support during this research work.

References

1. J. Kennedy, R.C. Eberhart, Particle swarm optimization, in *Proceedings of the IEEE International Conference on Neural Networks* (1995), pp. 1942–1948
2. J.H. Holland, Genetic algorithms. Sci. Am. (July 1992)
3. B. Akay, D. Karaboga, Artificial bee colony algorithm for large-scale problems and engineering design optimization. J. Intell. Manuf. **23**(4), 1001–1014 (2012)
4. D. Karaboga, B. Basturk, On the performance of artificial bee colony (ABC) algorithm. Appl. Soft Comput. **8**(1), 687–697 (2008). https://doi.org/10.1016/j.asoc.2007.05.007
5. M.A. Shinwan, L. Abualigah, M. Shehab, M.E.A. Elaziz, A. Khasawneh, H. Alabool, H. AlHamad, Dragonfly algorithm: a comprehensive survey of its results, variants, and applications. Multimed. Tools Appl. **80**, 14979–15016 (2021). https://doi.org/10.1007/s11042-020-10255-3
6. A. Kaveh, N. Farhoudi, A new optimization method: dolphin echolocation. Adv. Eng. Softw. **59**, 53–70 (2013). https://doi.org/10.1016/j.advengsoft.2013.03.004
7. L. Kn, B.R. Reddy, M.S. Kalavathi, Dolphin echolocation algorithm for solving optimal reactive power dispatch problem. Int. J. Comput. **12** (2014). https://doi.org/10.11648/j.ijepe.20140301.11
8. X.S. Yang, Chapter 10—Bat algorithms (2014). https://doi.org/10.1016/B978-0-12-416743-8.00010-5
9. Y.A. Alsariera, H.S. Alamri, A.M. Nasser, M.A. Majid, K.Z. Zamli, Comparative performance analysis of bat algorithm and bacterial foraging optimization algorithm using standard benchmark functions, in *2014 8th Malaysian Software Engineering Conference (MySEC)* (2014), pp. 295–300. https://doi.org/10.1109/MySec.2014.6986032
10. D. Merkle, M. Middendorf, Marco Dorigo and Thomas Stützle, Ant colony optimization, MIT Press (2004) ISBN 0-262-04219-3. Eur. J. Oper. Res. **168**(1), 269–271 (2006)
11. N.R. Haddaway, M.J. Page, C.C. Pritchard, L.A. McGuinness, PRISMA2020: an R package and Shiny app for producing PRISMA 2020-compliant flow diagrams, with interactivity for optimised digital transparency and Open Synthesis. Campbell Syst. Rev. **18**(2), e1230 (2022). https://doi.org/10.1002/cl2.1230
12. J.H. Holland, *Adaptation in Natural and Artificial Systems* (University of Michigan Press, Ann Arbor, MI, 1975)
13. M. Dorigo, Optimization, learning and natural algorithms (1992)
14. F. Glover, M. Laguna, *Tabu Search* (Kluwer Academic Publishers, Norwell, MA, USA, 1997)
15. D. Karaboga, An idea based on honey bee swarm for numerical optimization (2005)
16. K. Kaipa, D. Ghose, Glowworm swarm based optimization algorithm for multimodal functions with collective robotics applications. Multiagent Grid Syst. **2**, 209–222 (2006). https://doi.org/10.3233/MGS-2006-2301
17. S.C. Chu, P.W. Tsai, J.S. Pan, Cat swarm optimization (2006), pp. 854–858. https://doi.org/10.1007/11801603_94
18. X.S. Yang, S. Deb, Cuckoo search via Lévy flights, in *2009 World Congress on Nature & Biologically Inspired Computing (NaBIC)* (2009), pp. 210–214. https://doi.org/10.1109/NABIC.2009.5393690
19. X.S. Yang, Firefly algorithm, stochastic test functions and design optimisation. Int. J. Bio-Inspired Comput. **2** (2010). https://doi.org/10.1504/IJBIC.2010.032124
20. X.S. Yang, Bat algorithm: literature review and applications. Int. J. Bio-Inspired Comput. **5** (2013). https://doi.org/10.1504/IJBIC.2013.055093
21. T.Q. Wu, M. Yao, J.H. Yang, Dolphin swarm algorithm. Front. Inf. Technol. Electron. Eng. **17**, 717–729 (2016). https://doi.org/10.1631/FITEE.1500287
22. C. Caraveo, F. Valdez, O. Castillo, A new optimization metaheuristic based on the self-defense techniques of natural plants applied to the CEC 2015 benchmark functions, in *Advances in Fuzzy Logic and Technology 2017—Proceedings of: {EUSFLAT-2017}—The 10th Conference of the European Society for Fuzzy Logic and Technology, September 11–15, 2017, Warsaw,*

Poland IWIFSGN'2017—The Sixteenth International Workshop on Intuitionistic Fuzzy Sets and Generalized Nets, September 13–15, 2017, Warsaw, Poland, Volume 1, ed. J. Kacprzyk, E. Szmidt, S. Zadrozny, K.T. Atanassov, M. Krawczak, vol. 641 (Springer, 2017), pp. 380–388. https://doi.org/10.1007/978-3-319-66830-7_34

23. F. Martínez-Álvarez, G. Asencio-Cortés, J.F. Torres, D. Gutiérrez-Avilés, L. Melgar-García, R.P érez-Chacón, C. Rubio-Escudero, J.C. Riquelme, A. Troncoso (2020)
24. H. Carreon-Ortiz, F. Valdez, O. Castillo, A new discrete mycorrhiza optimization nature-inspired algorithm. Axioms **11**(8), 391 (2022). https://doi.org/10.3390/AXIOMS11080391
25. S.A. Ansari, A. Zafar, A fusion of dolphin swarm optimization and improved sine cosine algorithm for automatic detection and classification of objects from surveillance videos. Measurement **192**, 110921 (2022). https://doi.org/10.1016/j.measurement.2022.110921
26. J. Wu, M. Khishe, M. Mohammadi, S.H.T. Karim, M. Shams, Acoustic detection and recognition of dolphins using swarm intelligence neural networks. Scopus **115** (2021). https://doi.org/10.1016/j.apor.2021.102837
27. S. Sankar Ganesh, S. Rajaprakash, Dolphin swarm optimization algorithm for software-defined antenna selection algorithm in underwater acoustic sensor network. Int. J. Commun. Syst. **34** (2021). https://doi.org/10.1002/dac.4903
28. B.S. Mostafa, F.A. Alsalman, Application project task scheduling using dolphin swarm technology. Scopus **23**(1), 549–557 (2021). https://doi.org/10.11591/ijeecs.v23.i1.pp549-557
29. N.U. Reddy, Reducing the network latency to maintain network stability in UASN by using bio-inspired algorithms (2023). https://doi.org/10.1109/ICAISS58487.2023.10250497
30. A. Kashiv, H.K. Verma, Dolphin echolocation algorithm for small-signal stability analysis of DFIG-based wind power system. Scopus **12**(4), 301–328 (2021)
31. A.A. Fadhil, R.G.H. Alsarraj, A.M. Altaie, Software cost estimation based on dolphin algorithm. Scopus **8**, 75279–75287 (2020). https://doi.org/10.1109/ACCESS.2020.2988867
32. W. Qiao, Z. Yang, An improved dolphin swarm algorithm based on Kernel Fuzzy C-means in the application of solving the optimal problems of large-scale function. Scopus **8**, 2073–2089 (2020). https://doi.org/10.1109/ACCESS.2019.2958456
33. W. Qiao, Z. Yang, Modified dolphin swarm algorithm based on chaotic maps for solving high-dimensional function optimization problems. Scopus **7**, 110472–110486 (2019). https://doi.org/10.1109/ACCESS.2019.2931910
34. Y. Li, X. Wang, Improved dolphin swarm optimization algorithm based on information entropy. Scopus **67**(4), 679–685 (2019). https://doi.org/10.24425/bpasts.2019.130177
35. S. Amic, K.M.S. Soyjaudah, G. Ramsawock, Dolphin swarm algorithm for cryptanalysis. Scopus **863**, 149–163 (2019). https://doi.org/10.1007/978-981-13-3338-5_15
36. S. Sharma, A. Kaul, Hybrid fuzzy multi-criteria decision making based multi cluster head dolphin swarm optimized IDS for VANET. Scopus **12**, 23–38 (2018). https://doi.org/10.1016/j.vehcom.2017.12.003
37. F. Valdez, O. Castillo, P. Melin, Bio-inspired algorithms and its applications for optimization in fuzzy clustering. Algorithms **14**, 122 (2021). https://doi.org/10.3390/a14040122
38. N.J. Van Eck, L. Waltman, Software survey: VOSviewer, a computer program for bibliometric mapping. Scientometrics **84**(2), 523–538 (2010)
39. F. Valdez, J.C. Vazquez, P. Melin, O. Castillo, Comparative study of the use of fuzzy logic in improving particle swarm optimization variants for mathematical functions using co-evolution. Appl. Soft Comput. **52**, 1070–1083 (2017)
40. D. Sanchez, P. Melin, O. Castillo, A grey wolf optimizer for modular granular neural networks for human recognition. Comput. Intell. Neurosci. **2017** (2017). https://doi.org/10.1155/2017/4180510
41. O. Castillo, E. Lizzarraga, J. Soria, P. Melin, F. Valdez, New approach using ant colony optimization with ant set partition for fuzzy control design applied to the ball and beam system. Inf. Sci. **294**, 203–215 (2015)
42. L. Amador-Angulo, O. Mendoza, J.R. Castro, A. Rodriguez-Diaz, P. Melin, O. Castillo, Fuzzy sets in dynamic adaptation of parameters of a bee colony optimization for controlling the trajectory of an autonomous mobile robot. Sensors **16**(9), 1458 (2016)

Optimal Fuzzy Logic Controller for DC Motor Using Grasshopper Optimization Algorithm

Camilo Caraveo, Leticia Cervantes, Jesús Soto, and Oscar Castillo

1 Introduction

Today there is a great social demand for devices or tools that operate using DC type energy that can be stored and transported from one place to another and the most common means are batteries, then the stored energy is used to activate or start different mechanisms such as; electric cars [1, 2], hand tools [3], robots for the home [4, 5], explorer robots [6, 7], air vehicles or drones [8, 9] to mention the most relevant. However, the authors of this work consider it is important to mention that DC motors are inside of any device that has mechanical mobility in order to perform any action.

Based on the bibliographic analysis realized; we can affirm that the social demand for acquiring an electrical mechanical device to satisfy or automate some tasks or processes are increasing, this forces that the electric motor industry to invest more in the research and production of motors to satisfy the expectations of the consumers to offer greater durability, performance, quality, a minimum of electrical consumption and a long useful life in operation.

In the literature there are many works related to DC motors [10, 11], in all the works they mention the implementation of different methods and techniques in order to improve the existing results, for example the use of PI, PID controllers, improvements

C. Caraveo · L. Cervantes (✉) · J. Soto
Autonomous University of Baja California, Tijuana, Baja California, Mexico
e-mail: leticia.cervantes.huerta@uabc.edu.mx

C. Caraveo
e-mail: camilo.caraveo@uabc.edu.mx

J. Soto
e-mail: soto.jesus@uabc.edu.mx

O. Castillo
Tijuana Institute of Technology, Tijuana, Baja California, Mexico
e-mail: ocastillo@tectijuana.mx

P. Melin and O. Castillo (eds.), *New Directions on Hybrid Intelligent Systems Based on Neural Networks, Fuzzy Logic, and Optimization Algorithms*, Studies in Computational Intelligence 1146, https://doi.org/10.1007/978-3-031-53713-4_14

in windings, different compounds of materials for its manufacture, different types of batteries such as nickel-hydride (Ni-MH), lead acid, lithium ions (Li-ION), lithium polymer (LiPo), to mention the most popular. Today, the combination of motors with bushel technology with lithium-ion batteries has been achieved, however we consider that the existing results can be improved by using other techniques.

The use of intelligent techniques for solving and optimizing problems have made great contributions to science, which benefits growth and technological advancement, specifically the electric motor industry and tool manufacturers that use this type of motor. This paper presents the implementation of an optimization algorithm bioinspired by the behavior of grasshoppers in nature [3], this algorithm is used to optimize the parameters of the membership functions of a Mamdani-type fuzzy controller to control the speed of a DC motor [2, 11, 12].

2 Background of the Proposal

In the literature there are many published works about bio-inspired optimization algorithms, for example; Cuckoo search optimization [13], whale algorithm [14], antlion algorithms [15], gray wolf algorithm [16], bee colony optimization [17] to mention the most relevant. It is important to mention that the above-mentioned algorithms are in artificial intelligence, in the category of collective intelligence. We can also find multiple implementations to solve optimization problems in engineering, as well as combinations with other artificial intelligence techniques such as artificial neural networks, machine learning and deep learning to mention a few [18, 19].

This section present some works related to our work proposal for this manuscript, at present; fuzzy control has been implemented in multiple electronic devices, for example: in the automotive industry to control ABS braking [20], heated seats, hydraulic steering, gearbox [21], et al. Fuzzy logic has also been applied to devices for the home, for example; Microwave ovens, washing machines, refrigerators, smart vacuum cleaners, heating system, smart and automatic gardens [22], smart houses [23]. Other implementations are also mentioned, such as: cameras [24], smart televisions [25] to improve the images taken with smartphone cameras [26], to mention the most relevant.

As we can see in the previous paragraph, fuzzy logic has provided great contributions to science through its implementation to solve different problems that involve intelligent control. It is important to mention that the traditional way of calibrating and programming a controller was manually, that is trial and error until finding the optimal values and distribution for the membership functions. The performance of the fuzzy controller depends on these values, in recent years a new technique was developed to determine these optimal values for the controller [27], to that search and optimization algorithms are implemented. This technique has achieved the reduction of time in determining the optimal values, with a relatively low computational cost.

Bio-inspired optimization algorithms belong to the category of collective artificial intelligence, and have achieved great success in solving engineering problems, for example; solving systems of complex equations [28], optimization of hyperparameters for machine learning models [29], optimization of PID proportional derivative controllers [30], resolution of problems involving time series and artificial neural networks [31], as we can see In the previous paragraphs, the combination of heuristic or bio-inspired algorithms with deep learning models and fuzzy logic have managed to obtain favorable results for industry and science.

3 Proposed Method

This section describes the characteristics of the proposed optimization algorithm and the modifications made to adapt it to the problem to be solved in this work, also the characteristics of the implemented fuzzy logic controller are described.

3.1 Grasshopper Optimization Algorithm GOA

The GOA is a bio-inspired optimization algorithm based on the behavior and interaction of grasshopper swarms in nature, also is a population-based algorithm. This algorithm is a recent creation proposed by Saremi and Mirjalili in 2017 [32]. In this short lifetime of the algorithm, it has managed to prove that it is efficient and stable, as well as adaptation flexibility to work together with other intelligent methods or techniques. On the official website of the author (https://seyedalimirjalili.com/goa) [33], we can find three variants of this same algorithm, multi-objective optimization [34], Single-objective optimization [35] and the most recent version is a fast version of GOA, based on the analysis of those different variants, it is proposed to test and implement the fast version of the algorithm with the objective of avoiding the computational cost [32] and the main characteristics are described below.

Grasshoppers are insects and they are considered a pest due to their damage to crop production, agriculture and forestry in general. The life cycle of grasshoppers is illustrated in Fig. 1. Although grasshoppers are generally seen singly in the wild, however, they join a swarm, according to experts [36]. There is no limit to the size of the population, so it can be catastrophic or a nightmare for farmers, the population or swarm can be made up of adult grasshoppers and nymphs or both and their behavior is similar, they eat and devour almost all the vegetation in its path on the long migration journey.

The life cycle of grasshoppers consists of the three phases:

Egg: the initial phase involves reproduction by laying eggs on the ground strategically near a plant or vegetation that serves as food for the grasshopper when it hatches. This stage is around a few weeks before winter. Then, as soon as the soil temperature rises, the egg keeps developing and hatches to become a nymph.

Nymph: At this stage the grasshopper does not yet have wings. However, the nymph has to go through six stages to become an adult, with the wing buds growing

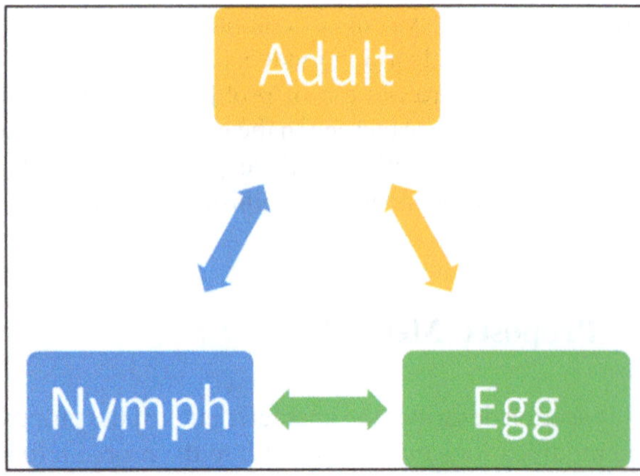

Fig. 1 Life cycle of grasshoppers

in size at each stage. Therefore, they move by walking or hopping while eating every vegetation in their path.

Adulthood: The wings expand and are fully functional after the stage they are ready to become part of a swarm and move quickly to a large-scale region, and start the breeding process.

These are the three stages of the grasshopper life cycle which are illustrated in Fig. 1.

The main characteristic of the swarm in the nymph phase is the slow movement and small steps of the grasshoppers. In contrast, long-range movement is the essential characteristic of the swarm in adulthood. These movements are used in the search process for new food sources, these characteristics are typical of algorithms inspired by nature, the logic of the search process is divided into two activities: exploration and exploitation. In exploration, search agents are forced to move within a search space, the space can be local or global. These two functions, in addition to the search for targets, are naturally performed by grasshoppers. Grasshopper swarm behavior is modeled using Eq. 1.

$$Xi = Si + Gi + Ai \tag{1}$$

where:

- Xi Defines the position of the i-th grasshopper.
- Si Represent the social interaction.
- Gi Represent the gravity force on the i-th grasshopper.
- Ai Represent the wind advection.

Note:

That to provide random behavior the Eq. 2 can be written as:

$$x_i = r_1 s_i + r_2 G_i + r_3 A_i \tag{2}$$

where:

r_1, r_2, and r_3 are random numbers in [0,1].

Firstly, we will start by describing the equation to force the social interaction Si between the grasshoppers, see Eq. 3.

$$s_i = \sum_{j=1}^{N} S(d_{ij})\widehat{d_{ij}},\ where\, i \neq j \tag{3}$$

where:

$d_{ij} = |x_j - x_i|$ Represent the distance between *i-th* grasshopper and the *j-th* grasshopper.

$\widehat{d_{ij}} = \frac{|x_j - x_i|}{d_{ij}}$ Represent a unit vector from the *i-th* grasshopper to the *j-th* grasshopper.

s Is a function to define the strength of social forces (repulsion and attraction between grasshoppers), as shown in Eq. 4.

$$s(r) = fe\frac{-r}{l} - e^{-r} \tag{4}$$

The attraction or repulsion is based on the following rule:

When the distance between two grasshoppers is between [0, 2.079], repulsion occurs, and when the distance between two grasshoppers is 2.079, neither attraction nor repulsion occurs, which form a comfort zone. When the distance exceeds [2.079], the attraction force increases, then progressively decreases until it reaches [4].

The function s fails to apply forces between grasshoppers when the distance between them is larger than 10. In order to solve this problem, we map the distance of grasshoppers in the interval [1,4]. In Fig. 2 we can observe the behavior.

Continuing with the variables involved in Eq. 1, the equation that represents the calculation of the force of gravity G_I, is described below, see Eq. 5.

$$G_I = -\widehat{ge_g} \tag{5}$$

where:

$-g$ represents the gravitational constant.

$\widehat{ge_g}$ Represent a unit vector toward center of earth.

The force and direction of the wind Ai, is determined by Eq. 6, the results affect to a great extent the nymphs and the adult grasshoppers. As a result, the movements of the nymphal and adult katydids are correlated with the wind direction Ai. See Eq. 6.

$$Ai = u\widehat{e_w} \tag{6}$$

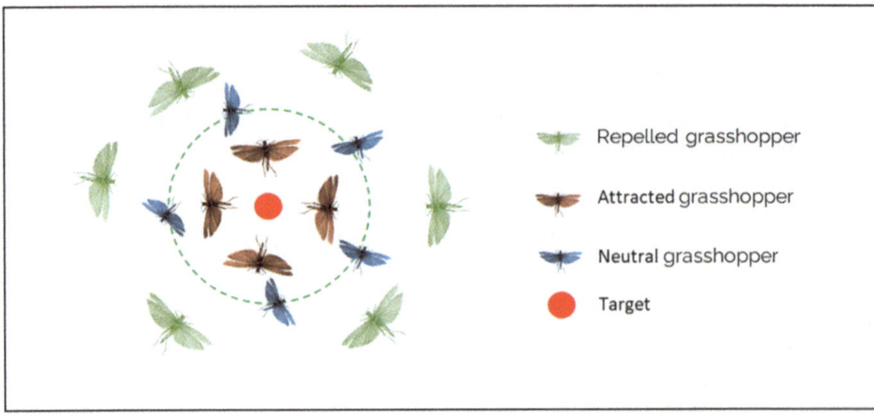

Fig. 2 Social interaction behavior

where u is a constant drift and $\widehat{e_w}$ is a unity vector in the direction of wind.

The position of the grasshoppers is determined by the following Eq. 7, in summary, it is the integration of the equations presented previously.

$$x_i = \sum_{j=1}^{N} S(d_{ij})\widehat{d_{ij}} - g\widehat{e_g} + u\widehat{e_w}$$
$$= \sum_{j=1}^{N} S(|x_j - x_i|)\frac{|x_j - x_i|}{d_{ij}} - g\widehat{e_g} + u\widehat{e_w},\ where\, i \neq j \quad (7)$$

In order to avoid premature convergence and stagnation in the search for solutions, the authors of this algorithm propose the following: this mathematical model cannot be used directly to solve optimization problems, mainly because the grasshoppers quickly reach the comfort zone and the swarm does not converge to a specified point. A modified version of this equation is proposed as follows to solve optimization problems, see Eq. 8.

$$x_i^d = c\left(\sum_{j=1}^{N} c\frac{UB_d - LB_d}{2} s(|x_j^d - x_i^d|)\frac{|x_j - x_i|}{d_{ij}}\right)$$
$$+ Best\, solution,\ where\, i \neq j \quad (8)$$

where $G = 0$, A is the best solution in the $d - th$ dimension, UB_d and LB_d are the upper and lower bounds in the $d - th$ dimension, respectively.

The parameter c represents the decreasing coefficient, and it is in charge of decreasing the comfort zone, repulsion zone, and attraction zone. In order to balance the exploration and the exploitation phases using the grasshopper approach, the coefficient c decreases according to the number of iterations. In the Eq. 9, we present a mathematical model to calculate the value c:

$$c = c_{max} - iter\frac{c_{max} - c_{min}}{Max_{iter}} \tag{9}$$

where c_{max} and c_{min} represent the maximum and minimum values of c respectively, $iter$ is the current iteration, and Max_{iter} is the maximum number of iterations.

The authors of this work consider it very important to inform the reader, the mathematical equations shown in the previous paragraphs, were taken from the technical paper developed by the creator of the grasshopper optimization algorithm.

3.2 Fuzzy Logic Controller Features

As already mentioned in the previous paragraphs of this manuscript, fuzzy logic has brought great contributions to science and engineering through optimization and problem solving, and we know that fuzzy logic is a discipline that is part of artificial intelligence. In this section the characteristics of the fuzzy controller of the problem proposed to optimize are described [37, 38], In Fig. 3, an illustration of the characteristics is presented.

The Fig. 3 shows the illustration of the default fuzzy controller of the model to be optimized, the characteristics are described below:

The controller is of the Mamdani type, has 2 input variables (error and error change), each input with 3 membership functions type trapezoidal and triangular in the center, in a distribution between [−1 1]. The 3 membership functions are distributed in three linguistic labels (negative, zero, positive), and a total of 9 if-the linguistic rules, with that number of rules, membership and type functions are enough

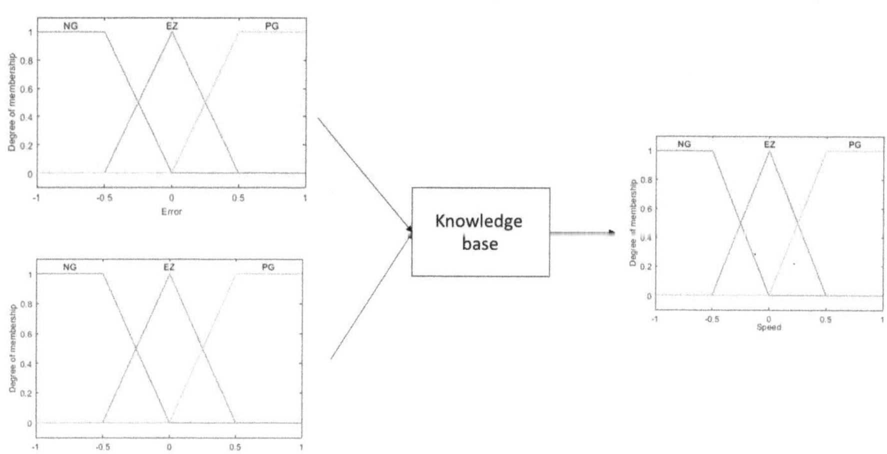

Fig. 3 Original fuzzy logic controller

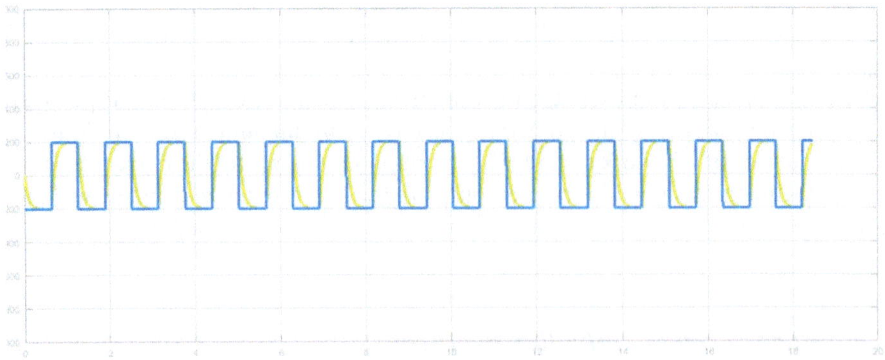

Fig. 4 Results from the original fuzzy controller

to solve those problems, the results and the behavior of the original fuzzy controller are presented in Fig. 4.

The results present an acceptable behavior, however the error is large for the magnitude of the problem, this error can generate a high cost, if this controller were implemented in an electric vehicle, the performance would be low in response to the acceleration demand of the user or driver.

Where the blue line represents the reference; i.e. the desired value, and the yellow line represents the output of the original controller, As can be seen, the results are not favorable.

3.3 Proposed Fuzzy Logic Controller

The main objective in this work is the implementation of the grasshopper algorithm for the optimization of fuzzy systems applied to control the speed of a DC motor. In Fig. 5 we present a general scheme of the proposal and contribution for this work.

The steps performed in the process of optimization and search for values for the membership functions are described below:

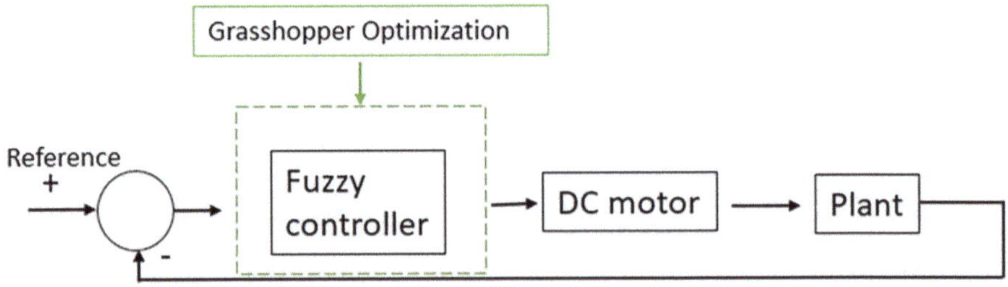

Fig. 5 Optimization fuzzy controller scheme

1. The first step is to select and analyze the characteristics of the problem to optimize, these data are described in Figs. 3 and 4.
2. The optimal range of search space for the values of the variables for the membership functions is determined.
3. The search ranges are encoded in the optimization algorithm, in each iteration the values for the membership function are adjusted in real time and the performance of the controller is evaluated using a virtual model.
4. The execution of the algorithm ends when the values found for the membership functions give an optimal result.

The base fuzzy system that is taken to be optimized is shown in Fig. 4. And the rules of the fuzzy system are presented in Table 1.

The problems that use fuzzy systems have shown great improvements, but sometimes the fuzzy systems need to be adjusted for optimal behavior in their system, which is why in this research the optimization of the fuzzy system is carried out using the grasshopper algorithm (see Fig. 4).

Table 2 shows the results obtained for 30 experiments carried out and the configuration of the basic parameters of the optimization algorithm.

As we can see in Table 2, the results obtained for the experiments carried out are presented, the average and the standard deviation of all the experiments are also shown, It is observed that at the beginning of the iterations the fast algorithm performs a global exploration in order to identify an optimal search space, after determining the search space, the algorithm performs a local convergence and focuses on finding the solution for the problem in only that zone, probably because the search space is restricted and the optimal solution may be found in a wider area, it may be an improvement for a new variant of the algorithm.

Table 1 Rules of the fuzzy system

Error	Change_error	Output
NG	NG	NG
NG	EZ	NG
NG	PG	EZ
EZ	NG	NG
EZ	EZ	EZ
EZ	PG	PG
PG	NG	EZ
PG	EZ	PG
PG	PG	PG

Table 2 Results of the simulations

Num	Num. agents	Num. inter	Best value	Worst value
1	15	20	15.2227	15.3194
2	20	15	10.8839	11.4615
3	15	25	10.8839	10.8976
4	15	20	10.8839	11.4615
5	15	12	13.684	13.8715
6	20	15	10.8839	11.4615
7	20	15	10.9137	10.9275
8	20	15	20.37185	98.3495
9	20	15	13.0304	15.7234
10	20	15	21.1376	201.9147
11	20	15	13.9182	26.2968
12	20	15	12.11016	12.9949
13	20	15	14.5632	25.3102
14	20	15	11.8848	96.2877
15	20	15	10.8946	12.6987
		Average	13.41779	38.33176
		Standard deviation	3.331395	53.87981

4 Conclusions

In this work, the integration of the bio-inspired algorithm in the behavior of grasshoppers in nature to a fuzzy logic controller was presented, for the tuning of optimal values of the membership functions, and the controller was applied to optimize the performance of a DC motor, using the original controller of this model as a reference, we can conclude that the results found exceeded the expected expectations, because it is a complex problem and the algorithm is new, however it is a good contribution to science to consider this algorithm in future works. Derived from this implementation we can say that the algorithm behaved stable and with acceptable performance, however, there are some improvements that can be made in order to improve the results, one improvement is; give it more exploration time, in order to consider other search spaces and avoid premature convergence, stagnation in local minimum. However, the results found were better than the original version of the proposed model, therefore we can conclude that the proposal exceeds expectations. As future work, we could consider other optimization metaheuristics, such as [39–42].

References

1. E. Agamloh, A. von Jouanne, A. Yokochi, An overview of electric machine trends in modern electric vehicles. Machines **8**(2) (2020). https://doi.org/10.3390/MACHINES8020020
2. C. Huang, F. Lei, X. Han, Z. Zhang, Determination of modeling parameters for a brushless DC motor that satisfies the power performance of an electric vehicle. Meas. Control (U.K.) **52**(7–8) (2019). https://doi.org/10.1177/0020294019842607
3. Y. Meraihi, A.B. Gabis, S. Mirjalili, A. Ramdane-Cherif, Grasshopper optimization algorithm: theory, variants, and applications. IEEE Access **9** (2021). https://doi.org/10.1109/ACCESS.2021.3067597
4. C. Chen, T. Zhang, A review of design and fabrication of the bionic flapping wing micro air vehicles. Micromachines **10**(2) (2019). https://doi.org/10.3390/mi10020144
5. J.R. Ruiz-Sarmiento, C. Galindo, J. Gonzalez-Jimenez, Robot@Home, a robotic dataset for semantic mapping of home environments. Int. J. Rob. Res. **36**(2) (2017). https://doi.org/10.1177/0278364917695640
6. J. Sundin, K. Kokmanian, M.K. Fu, S. Bagheri, M. Hultmark, A soft material flow sensor for micro air vehicles. Soft Robot. **8**(2) (2021). https://doi.org/10.1089/soro.2019.0130
7. A. Mueller, Modern robotics: mechanics, planning, and control [bookshelf]. IEEE Control Syst. **39**(6) (2022). https://doi.org/10.1109/mcs.2019.2937265
8. M.J. Mahmoodabadi, N. Rezaee Babak, Fuzzy adaptive robust proportional–integral–derivative control optimized by the multi-objective grasshopper optimization algorithm for a nonlinear quadrotor. JVC/J. Vib. Control **26**(17–18) (2020). https://doi.org/10.1177/1077546319901019
9. S. Dwivedi, Detecting anonymous attacks in wireless communication medium using adaptive grasshopper optimization algorithm. Cogn. Syst. Res. **69** (2021). https://doi.org/10.1016/j.cogsys.2021.04.003
10. J. Baranowski, T. Drabek, P. Piątek, A. Tutaj, Diagnosis and mitigation of electromagnetic interference generated by a brushless dc motor drive of an electric torque tool. Energies **14**(8) (2021). https://doi.org/10.3390/en14082149
11. F. Naseri, E. Farjah, E. Schaltz, K. Lu, N. Tashakor, Predictive control of low-cost three-phase four-switch inverter-fed drives for brushless DC motor applications. IEEE Trans. Circuits Syst. I Regul. Pap. **68**(3) (2021). https://doi.org/10.1109/TCSI.2020.3043468
12. S. Hassan, M.A. Khanesar, N.K. Hussein, S.B. Belhaouari, U. Amjad, W.K. Mashwani, Optimization of interval type-2 fuzzy logic system using grasshopper optimization algorithm. Comput. Mater. Contin. **71**(2) (2022). https://doi.org/10.32604/cmc.2022.022018
13. C.E.M. Barbosa, G.C. Vasconcelos, Cuckoo search optimization for short term wind energy forecasting (2016). https://doi.org/10.1109/CEC.2016.7744002
14. A. Kaveh, M.R. Moghaddam, A hybrid WOA-CBO algorithm for construction site layout planning problem. Sci. Iran. **25**(3A) (2018). https://doi.org/10.24200/sci.2017.4212
15. L. Abualigah, M. Shehab, M. Alshinwan, S. Mirjalili, M.A. Elaziz, Ant lion optimizer: a comprehensive survey of its variants and applications. Arch. Comput. Methods Eng. **28**(3) (2021). https://doi.org/10.1007/s11831-020-09420-6
16. M.H. Nadimi-Shahraki, S. Taghian, S. Mirjalili, An improved grey wolf optimizer for solving engineering problems. Expert Syst. Appl. **166** (2021). https://doi.org/10.1016/j.eswa.2020.113917
17. Tehzeeb-ul-Hassan, T. Alquthami, S.E. Butt, M.F. Tahir, K. Mehmood, Short-term optimal scheduling of hydro-thermal power plants using artificial bee colony algorithm. Energy Rep. **6** (2020). https://doi.org/10.1016/j.egyr.2020.04.003
18. C. Ning, F. You, Optimization under uncertainty in the era of big data and deep learning: when machine learning meets mathematical programming. Comput. Chem. Eng. **125** (2019). https://doi.org/10.1016/j.compchemeng.2019.03.034
19. Y. Ma, R. Han, W. Wang, Portfolio optimization with return prediction using deep learning and machine learning. Expert Syst. Appl. **165** (2021). https://doi.org/10.1016/j.eswa.2020.113973

20. S. Latreche, S. Benaggoune, Robust wheel slip for vehicle anti-lock braking system with fuzzy sliding mode controller (FSMC). Eng. Technol. Appl. Sci. Res. **10**(5) (2020). https://doi.org/10.48084/etasr.3830
21. W. Batayneh, A. Bataineh, M.A. Jaradat, Intelligent adaptive fuzzy logic genetic algorithm controller for anti-lock braking system. Int. Rev. Model. Simul. **14**(1) (2021). https://doi.org/10.15866/iremos.v14i1.19838
22. F. Azmi, J. Louise, Z.R. Sitompul, S. Kumar, J. Surya, Design of smart garden sprinklers based on fuzzy logic. J. Inform. Telecommun. Eng. **4**(1) (2020). https://doi.org/10.31289/jite.v4i1.3886
23. A. Gozuoglu, O. Ozgonenel, S. Karagol, Fuzzy logic based low cost smart home application (2019). https://doi.org/10.23919/ELECO47770.2019.8990409
24. V. Korolyov, M. Ogurtsov, A. Khodzinsky, Statement of the problem of complete set of UAV group on the basis of models of granular calculations and fuzzy logic. Cybern. Comput. Technol. (2) (2021). https://doi.org/10.34229/2707-451x.21.2.3
25. A. Yohandrik Bi Dawe, P. Dani Prasetyo Adi, Electricity monitoring system based on fuzzy logic and Internet of things. Nu **01** (2021)
26. M.K. Hasan, M.Z. Chowdhury, M. Shahjalal, Y.M. Jang, Fuzzy based network assignment and link-switching analysis in hybrid OCC/LiFi system. Wirel. Commun. Mob. Comput. **2018** (2018). https://doi.org/10.1155/2018/2870518
27. E. Bernal, O. Castillo, J. Soria, F. Valdez, Imperialist competitive algorithm with dynamic parameter adaptation using fuzzy logic applied to the optimization of mathematical functions. Algorithms **10**(1) (2017). https://doi.org/10.3390/a10010018
28. M. Abdel-Basset, R. Mohamed, M. Abouhawwash, Balanced multi-objective optimization algorithm using improvement based reference points approach. Swarm Evol. Comput. **60** (2021). https://doi.org/10.1016/j.swevo.2020.100791
29. L. Yang, A. Shami, On hyperparameter optimization of machine learning algorithms: theory and practice. Neurocomputing **415** (2020). https://doi.org/10.1016/j.neucom.2020.07.061
30. S. Biswas, P.K. Roy, K. Chatterjee, FACTS-based 3DOF-PID controller for LFC of renewable power system under deregulation using GOA. IETE J. Res. (2021). https://doi.org/10.1080/03772063.2020.1870874
31. H. Moayedi, A. Mosavi, Double-target based neural networks in predicting energy consumption in residential buildings. Energies **14**(5) (2021). https://doi.org/10.3390/en14051331
32. S. Saremi, S. Mirjalili, A. Lewis, Grasshopper optimisation algorithm: theory and application. Adv. Eng. Softw. **105** (2017)
33. Seyedali Mirjalili, SEYEDALI MIRJALILI 2019 (2022), https://seyedalimirjalili.com/goa
34. A. Tabak, Fractional order frequency proportional-integral-derivative control of microgrid consisting of renewable energy sources based on multi-objective grasshopper optimization algorithm. Trans. Inst. Meas. Control **44**(2) (2022). https://doi.org/10.1177/01423312211034660
35. A.A.S. Bahdad, S.F.S. Fadzil, H.O. Onubi, S.A. BenLasod, Sensitivity analysis linked to multi-objective optimization for adjustments of light-shelves design parameters in response to visual comfort and thermal energy performance. J. Build. Eng. **44** (2021). https://doi.org/10.1016/j.jobe.2021.102996
36. S.J. Simpson, A.R. McCaffery, B.F. HÄgele, A behavioural analysis of phase change in the desert locust. Biol. Rev. **74**(4) (1999). https://doi.org/10.1111/j.1469-185X.1999.tb00038.x
37. S. Sadi, DC motor speed control using Mamdani fuzzy logic based on microcontroller. J. Tek. **9**(2) (2020). https://doi.org/10.31000/jt.v9i2.3676
38. Y.A. Almatheel, A. Abdelrahman, Speed control of DC motor using fuzzy logic controller (2017). https://doi.org/10.1109/ICCCCEE.2017.7867673
39. F. Valdez, J.C. Vazquez, P. Melin, O. Castillo, Comparative study of the use of fuzzy logic in improving particle swarm optimization variants for mathematical functions using co-evolution. Appl. Soft Comput. **52**, 1070–1083 (2017)
40. O. Castillo, E. Lizzarraga, J. Soria, P. Melin, F. Valdez, New approach using ant colony optimization with ant set partition for fuzzy control design applied to the ball and beam system. Inf. Sci. **294**, 203–215 (2015)

41. L. Amador-Angulo, O. Mendoza, J.R. Castro, A. Rodriguez-Diaz, P. Melin, O. Castillo, Fuzzy sets in dynamic adaptation of parameters of a bee colony optimization for controlling the trajectory of an autonomous mobile robot. Sensors **16**(9), 1458 (2016)
42. D. Sanchez, P. Melin, O. Castillo, A grey wolf optimizer for modular granular neural networks for human recognition. Comput. Intell. Neurosci. **2017** (2017). https://doi.org/10.1155/2017/4180510

Parameter Exploration in the Artificial Gorilla Troops Optimizer Algorithm

Ivette Miramontes and Patricia Melin

1 Introduction

Currently, the world is immersed in an era of increasing complexity, where the challenges and problems we face are more diverse and multidimensional than ever before. In this context, the search for effective solutions has become a crucial task in a wide range of disciplines, from optimizing industrial processes to making decisions [1, 2] in the field of healthcare [3, 4], and solving logistics and transportation problems [5, 6], to name a few.

In the face of this increasing complexity, a diversified set of computational approaches has emerged to address these challenges. Among them, metaheuristics have stood out as a set of powerful tools that enable tackling complex problems and finding high-quality solutions. These techniques draw inspiration from natural processes and human heuristics to guide the search for optimal or near-optimal solutions in extensive search spaces [7–13]. Among the most current we have Starling Murmuration optimizer [14], Orca predation algorithm [15], Coati Optimization Algorithm [16], Artificial hummingbird algorithm [17], Coronavirus mask protection algorithm [18].

The main goal of this research is to conduct an investigation into the parameters of the Artificial Gorilla Troops Optimizer in order to observe its performance and study the outcome generated through diverse mathematical test functions. The aim is to ascertain the specific parameter combination at which the algorithm yields superior results.

The structure of this article has been organized as follows: in Sect. 2 the literature review is presented, in Sect. 3 the reference mathematical functions are described, the

I. Miramontes · P. Melin (✉)
Tijuana Institute of Technology, TecNM, Tijuana, BC, México
e-mail: pmelin@tectijuana.mx

P. Melin and O. Castillo (eds.), *New Directions on Hybrid Intelligent Systems Based on Neural Networks, Fuzzy Logic, and Optimization Algorithms*, Studies in Computational Intelligence 1146, https://doi.org/10.1007/978-3-031-53713-4_15

results are analyzed in Sect. 4 and, finally, the conclusions of this work are presented in Sect. 5.

2 Literature Review

This section describes the optimization method used, as well as works where this method has been applied to solve problems.

2.1 *Artificial Gorilla Troops Optimizer*

Abdollahzadeh [19] proposed the Artificial Gorilla Troops Optimizer method in 2021, inspired by the group behavior of gorillas. This method is based on five operators in the optimization phases of the algorithm, which are well-known as exploration and exploitation. For the case of exploration, it is proposed to use three operators based on the migration behaviors towards unknown places, significantly helping to increase the exploration of the problem provided, the second operator is based on moving towards other gorillas to have a balance between exploration and exploitation, and as the last operator of this phase we have to migrate to a known place, which increases the capacity of the method in the search on different optimization spaces. Regarding the exploitation phase, the operators to be used would be two which significantly improve the search performance in exploitation.

2.2 *Related Works Utilizing GTO Algorithm*

The popularity of the GTO has been increasing [20–24] and as time has passed it has become more used. Listed below are problems that have been optimized using this method:

In the Automatic Accurate Classification of Alzheimer [25], for Frequency Regulation of Wind Contributed Microgrid System [26], for Unmanned Aerial Vehicle (UAV)-assisted intelligent vehicular network [27], for the detection of distributed denial-of-service DDoS Attacks [28], which is based on a model of Machine Learning. Also, it is utilized in an Islanded Renewable Microgrid for optimizing Power Flow [29], and for Bearing Fault Diagnosis [30].

3 Mathematical Benchmark Functions

Classical benchmark mathematical functions are a series of mathematical functions that are commonly used in the optimization and evaluation of algorithms and numerical methods. These functions have become standard in the evaluation of optimization algorithms due to their simplicity, well-known properties, and ability to represent a variety of mathematical challenges [31]. The main characteristics of classical benchmark mathematical functions are:

Knowledge of the optimum: The optimal value of the function is known, which allows the performance of an algorithm to be evaluated based on its ability to find this optimum.

Variable dimensionality: These functions can be defined in multidimensional spaces, allowing algorithms to be evaluated in a variety of scenarios and dimensions.

Boundary conditions: They can include boundary conditions or constraints that represent real situations.

Diversity of behavior: Some benchmark functions are uni-modal, meaning that they have a single global minimum, while others are multimodal, having several local minima, which allows evaluating the ability of an algorithm to deal with different types of functions [31].

For this work, six unimodal Benchmark functions (F1–F6) are used, this means that they only have a single optimal value, and we selected this type due to the nature of the case study, it is only the analysis of the behavior of the results by varying the values of the algorithm parameters. The representations of the unimodal mathematical functions are provided in Table 1, and their graphical representations can be found in Fig. 1.

Table 1 Unimodal mathematical benchmark functions

No.	Function	Equation	Search space	Minin
F1	Sphere	$f(x) = \sum_{i=1}^{n_x} x_i^2$	[−100, 100]	0
F2	Schwefel 2.22	$f(x) = \sum_{i-1}^{n} \lvert x_i \rvert + \prod_{l=0}^{n} x_i$	[−10, 10]	0
F3	Schwefel 1.2	$f(x) = \sum_{i=1}^{n} \left(\sum_{j=1}^{i} x_j \right)^2$	[−100, 100]	0
F4	Schwefel 2.21	$f(x) = \max\{\lvert x_i \rvert, 1 \le i \le n\}$	[−100, 100]	0
F5	Rosenbrock	$f(x) = \sum_{i=1}^{n} \left[100(x_{i+1} - x_i^2)^2 + (x_i - 1)^2 \right]$	[−30, 30]	0
F6	Step	$f(x) = \sum_{i=1}^{n} \lfloor \lvert x_i \rvert \rfloor$	[−100, 100]	0

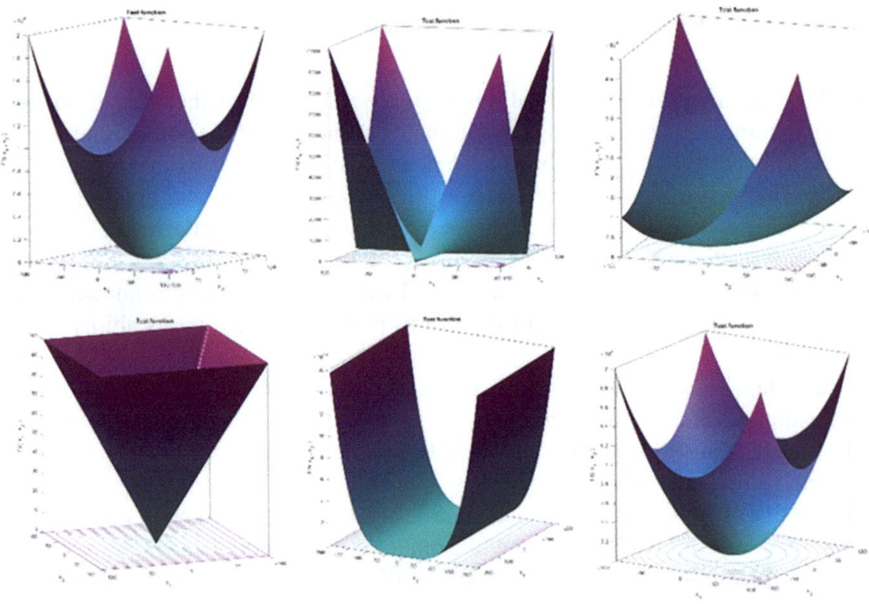

Fig. 1 Unimodal benchmark functions

4 Results

In this section, we carry out a comprehensive analysis of the parameters used in the GTO method, as described in the literature, which are composed of three fundamental elements. The parameter "p" plays a crucial role in determining the exploration of the algorithm, while "W" is mainly responsible for the exploitation of solutions, and "Beta" is a parameter that requires a prior assignment before the optimization operation. Within the framework of the research we present, a total of thirty independent experiments were conducted. In each experiment, 30 individuals were used for 500 iterations, using 50 dimensions. Figure 2 shows the graphic representation of the parameter analysis.

The initial values of the parameters used in the GTO algorithm were taken directly from the author's recommendations:

p: 0.03
W: 0.8
Beta: 3

Each of these parameters was modified independently to analyze their influence on the results obtained. Table 2 presents the results of the conducted experiments by varying the parameter "p" in a context of 50 dimensions. For this series, 30 independent experiments were carried out to examine how performance varied on six different mathematical functions. The "p" value was increased in steps of two

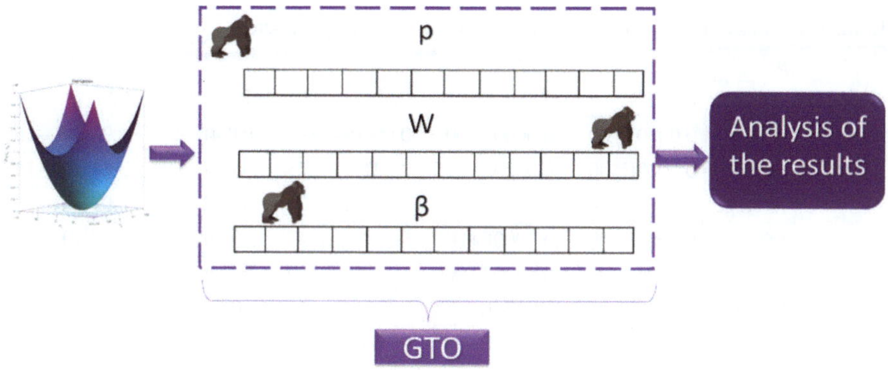

Fig. 2 Graphical representation of parameter analysis

units in each experiment. To provide a detailed analysis, both the best and worst values were obtained. In addition to that, the average and standard deviation of the results are presented.

Regarding the parameter "p", it is essential to understand its impact on the performance of the GTO algorithm. This parameter, which controls the algorithm's exploration probability, behaves interestingly depending on its value. By analyzing the functions F1 and F3 in detail, it is evident that optimal performance is achieved when "p" has low values. This means that the algorithm tends to explore the search space more intensely when "p" is small, which can be beneficial for finding high-quality solutions in contexts where meticulous exploration is crucial since according to the simulated behavior with low values it is exploring in unknown spaces.

However, as we increase the value of "p" above some critical threshold, usually around 0.5 in this case, there is a change in the behavior of the algorithm. Instead of exploring broadly, the algorithm begins to focus on the behavior of moving toward known solutions. This may be appropriate in some situations, but it can also lead to premature convergence, where the algorithm gets stuck in a local minimum without exploring other possible optimal solutions.

A similar phenomenon is observed in the function F4, where the optimal value and the average performance vary significantly as we modify "p". This underlines the importance of carefully tuning the parameter "p" based on the specific characteristics of the problem and the optimization goals. In summary, a small value of "p" promotes diversified exploration, while a higher value favors the exploration of known solutions. The choice of the optimal value of "p" depends on the nature of the problem and the desired strategy to find effective solutions.

Table 3 shows the detailed results of the analysis carried out on the "W" parameter. This parameter was adjusted incrementally, increasing its value by two units in each experiment. It is important to highlight that these experiments were carried out in a 50-dimensional context.

Concerning the parameter "W", the influence it exerts on the results is notable, especially when we observe both the best values and the averages in the functions

Table 2 Results obtained with GTO algorithms varying "p" parameter

	p					
Function		0.1	0.3	0.5	0.7	0.9
F1	Best	0.0000E+00	0.0000E+00	0.0000E+00	6.0424E−245	2.2469E−167
	Worst	0.0000E+00	0.0000E+00	2.6222E−276	5.9954E−217	3.0530E−152
	Average	0.0000E+00	0.0000E+00	8.8623E−278	1.9985E−218	2.0195E−153
	SD	0.0000E+00	0.0000E+00	0.0000E+00	0.0000E+00	7.1213E−153
F2	Best	2.9716E−199	0.0000E+00	5.2890E−150	7.8109E−121	1.4045E−90
	Worst	7.8088E−183	1.9774E−307	7.8979E−136	8.1069E−112	4.5571E−78
	Average	2.6084E−184	0.0000E+00	2.6895E−137	2.9942E−113	2.6694E−79
	SD	0.0000E+00	0.0000E+00	1.4411E−136	1.4770E−112	9.4926E−79
F3	Best	0.0000E+00	0.0000E+00	3.6634E−280	1.6291E−224	1.0732E−151
	Worst	0.0000E+00	1.9774E−307	2.7360E−250	7.1656E−201	9.8446E−134
	Average	0.0000E+00	0.0000E+00	9.1201E−252	2.4103E−202	4.2528E−135
	SD	0.0000E+00	0.0000E+00	0.0000E+00	0.0000E+00	1.8071E−134
F4	Best	8.7827E−203	2.0711E−177	1.1002E−150	5.8065E−120	1.2232E−84
	Worst	5.2641E−185	2.0903E−161	1.9344E−133	5.0098E−109	1.2092E−70
	Average	2.5519E−186	7.1533E−163	6.4491E−135	4.0411E−110	4.0847E−72
	SD	0.0000E+00	0.0000E+00	3.5317E−134	1.1811E−109	2.2068E−71
F5	Best	9.9846E−06	4.2231E−08	1.7892E−06	5.8789E−07	5.5316E−07
	Worst	3.8505E−02	1.4688E−02	6.7044E−03	6.3845E−03	2.3978E−03
	Average	4.2078E−03	1.1448E−03	1.1132E−03	7.7212E−04	3.9776E−04
	SD	8.2876E−03	2.7798E−03	1.6174E−03	1.2705E−03	6.3040E−04
F6	Best	1.0387E−08	3.7387E−08	2.2368E−08	1.3723E−10	6.2043E−08
	Worst	7.5798E−05	1.1041E−04	1.4268E−04	2.2073E−04	7.8413E−04
	Average	2.2758E−05	1.6687E−05	1.7410E−05	2.8068E−05	9.8881E−05
	SD	2.2302E−05	2.6893E−05	3.1805E−05	5.6798E−05	1.7468E−04

F1–F4 and F6. It is perceived that, by assigning small values to this parameter, the expected results are not obtained. This implies that the GTO algorithm does not favor exploitation when "W" is low.

It should be noted that as we gradually increase the value of "W", the results tend to improve. A larger value of "W" tilts the balance toward exploiting better solutions, which can be beneficial when a promising solution has been found. This means that, as "W" increases, the algorithm focuses more on refining already existing solutions, which can lead to faster convergence toward optimal solutions, especially in problems where excessive exploration is not necessary.

Table 4 reflects the results obtained by varying the Beta parameter. The purpose of this analysis lies in exploring the behavior of said parameter, increasing it by two units in each experiment. From a total of 30 independent experiments, both the best and worst values are presented, as well as the average and standard deviation.

Table 3 Results obtained with GTO algorithms varying W parameter

	W					
Function		0.1	0.3	0.5	0.7	0.9
F1	Best	0.0000E+00	0.0000E+00	0.0000E+00	0.0000E+00	0.0000E+00
	Worst	4.2345E−283	0.0000E+00	0.0000E+00	0.0000E+00	0.0000E+00
	Average	1.4117E−284	0.0000E+00	0.0000E+00	0.0000E+00	0.0000E+00
	SD	0.0000E+00	0.0000E+00	0.0000E+00	0.0000E+00	0.0000E+00
F2	Best	7.0960E−151	2.5679E−167	4.9133E−187	1.7322E−200	3.4693E−213
	Worst	3.7310E−139	1.2150E−153	8.8397E−171	3.5449E−183	3.5447E−200
	Average	1.6789E−140	5.7024E−155	3.0954E−172	1.1819E−184	1.6318E−201
	SD	7.1104E−140	2.2995E−154	0.0000E+00	0.0000E+00	0.0000E+00
F3	Best	6.3617E−301	0.0000E+00	0.0000E+00	0.0000E+00	0.0000E+00
	Worst	5.2308E−269	4.2041E−302	0.0000E+00	0.0000E+00	0.0000E+00
	Average	1.9823E−270	1.7126E−303	0.0000E+00	0.0000E+00	0.0000E+00
	SD	0.0000E+00	0.0000E+00	0.0000E+00	0.0000E+00	0.0000E+00
F4	Best	9.6577E−153	1.6314E−170	9.6680E−188	9.5099E−198	2.8903E−213
	Worst	3.3349E−141	8.3279E−154	4.1908E−171	8.9170E−184	2.1536E−193
	Average	2.0729E−142	2.8307E−155	1.4510E−172	3.0354E−185	1.2941E−194
	SD	7.0859E−142	1.4939E−154	0.0000E+00	0.0000E+00	0.0000E+00
F5	Best	1.5638E−05	1.9623E−04	9.4247E−07	2.7652E−04	3.8596E−06
	Worst	4.8346E+01	4.6479E+01	4.5803E+01	4.5628E+01	4.5826E+01
	Average	6.5640E+00	3.0955E+00	7.5424E+00	1.9464E+01	1.7845E+01
	SD	1.6381E+01	1.1714E+01	1.7099E+01	2.2627E+01	2.2225E+01
F6	Best	7.0068E−07	6.2936E−05	1.9699E−05	2.7652E−04	4.1916E−06
	Worst	3.4063E−01	2.5141E−02	2.2674E−01	4.5628E+01	1.3844E−03
	Average	5.5192E−02	5.9190E−03	8.5980E−03	1.9464E+01	2.1532E−04
	SD	6.7436E−02	5.4067E−03	4.1221E−02	2.2627E+01	3.4148E−04

In the results, an interesting trend stands out about the functions F2, F4, F5 and F6 as the Beta parameter increases. In the case of functions F2 and F4, it can be analyzed that as said parameter increases, the average value obtained decreases. This suggests that an increase in the Beta value may not be beneficial for these functions as it could lead to suboptimal results.

On the contrary, when the functions F5 and F6 are analyzed, we notice that, on average, the results remain relatively stable as Beta varies. This may mean that these functions are less sensitive to changes in Beta, which could be an indication that exploration and exploitation within the GTO algorithm have less impact on its performance.

These observations highlight the importance of understanding how different parameters affect the functions of interest in our study. Likewise, they suggest that

Table 4 Results obtained with GTO algorithms varying β parameter

	β					
Function		1	3	5	7	9
F1	Best	0.0000E+00	0.0000E+00	0.0000E+00	0.0000E+00	0.0000E+00
	Worst	0.0000E+00	0.0000E+00	0.0000E+00	0.0000E+00	0.0000E+00
	Average	0.0000E+00	0.0000E+00	0.0000E+00	0.0000E+00	0.0000E+00
	SD	0.0000E+00	0.0000E+00	0.0000E+00	0.0000E+00	0.0000E+00
F2	Best	1.6378E−283	1.5710E−204	1.0221E−186	6.7661E−175	3.1434E−166
	Worst	1.8745E−263	1.2874E−189	2.6841E−170	3.6463E−163	6.9835E−158
	Average	6.2523E−265	4.2916E−191	9.3935E−172	1.3821E−164	4.5020E−159
	SD	0.0000E+00	0.0000E+00	0.0000E+00	0.0000E+00	0.0000E+00
F3	Best	0.0000E+00	0.0000E+00	0.0000E+00	0.0000E+00	0.0000E+00
	Worst	0.0000E+00	0.0000E+00	0.0000E+00	2.6674E−302	4.8625E−290
	Average	0.0000E+00	0.0000E+00	0.0000E+00	8.8991E−304	2.7168E−291
	SD	0.0000E+00	0.0000E+00	0.0000E+00	0.0000E+00	0.0000E+00
F4	Best	1.4345E−277	2.3799E−202	1.1149E−185	1.3392E−172	7.0697E−167
	Worst	2.5324E−259	3.9827E−189	1.2683E−167	9.4058E−161	1.1835E−155
	Average	9.9714E−261	1.3405E−190	4.2393E−169	4.1263E−162	5.7843E−157
	SD	0.0000E+00	0.0000E+00	0.0000E+00	0.0000E+00	0.0000E+00
F5	Best	2.3309E−04	1.0657E−04	6.7829E−09	1.3541E−06	3.5322E−04
	Worst	4.7481E+01	4.5507E+01	4.5859E+01	4.7560E+01	4.5998E+01
	Average	2.0164E+01	2.2264E+01	1.2050E+01	1.6805E+01	1.6641E+01
	SD	2.3417E+01	2.2620E+01	2.0309E+01	2.2382E+01	2.2206E+01
F6	Best	1.6686E−04	9.6379E−06	1.2562E−05	1.4722E−05	5.6513E−08
	Worst	3.4201E−02	1.3052E−03	7.2633E−03	9.9547E−03	5.2772E−03
	Average	8.1472E−03	1.8671E−04	7.4729E−04	1.4140E−03	1.2297E−03
	SD	8.4297E−03	2.4757E−04	1.4061E−03	2.5832E−03	1.4362E−03

the adjustment of Beta should be carefully considered based on the specific goals and characteristics of the problem being addressed.

5 Conclusions and Future Work

This work presents an exhaustive analysis of the parameters of the algorithm called "Artificial Gorilla Troops Optimizer," specifically the parameters "p," "W," and "Beta," which play a crucial role in controlling exploration and exploitation in the said algorithm. To carry out this study, the values of these parameters were varied and their impact on the results was observed. Six unimodal Benchmark functions were used as test cases in the algorithm.

It was observed that, in the case of the parameter "p," optimal results were obtained when small values were used, while in the case of "W," a significant improvement in the results was observed when its value increased. Regarding "Beta," it was noted that in certain functions the results were improved when using small values, while in more complex functions such as F5 and F6, better results were obtained with a larger value.

As a possible line of future work, the implementation of dynamic parameter tuning in the GTO algorithm is considered, especially to solve more complex mathematical problems. In addition, we seek to explore the fuzzy application of this algorithm in fields such as fuzzy control [32–34] and medical diagnosis, which would significantly expand its application potential in various areas [35, 36]. Finally, we could try a similar approach with other metaheuristics, such as [37–40].

References

1. C. Li, Y. Chen, Y. Shang, A review of industrial big data for decision making in intelligent manufacturing. Eng. Sci. Technol., Int. J. **29**, 101021 (2022)
2. A. Zeiser, B. van Stein, T. Bäck, Requirements towards optimizing analytics in industrial processes. Procedia Comput. Sci. **184**, 597–605 (2021)
3. I. Miramontes, P. Melin, Interval type-2 fuzzy approach for dynamic parameter adaptation in the bird swarm algorithm for the optimization of fuzzy medical classifier. Axioms **11**(9), 1–29 (2022)
4. A.M. Anter, M. Ali, Feature selection strategy based on hybrid crow search optimization algorithm integrated with chaos theory and fuzzy c-means algorithm for medical diagnosis problems. Soft. Comput. **24**(3), 1565–1584 (2020)
5. K. Leng, S. Li, Distribution path optimization for intelligent logistics vehicles of urban rail transportation using VRP optimization model. IEEE Trans. Intell. Transp. Syst. **23**(2), 1661–1669 (2022)
6. H. Liu, P. Zhan, M. Zhou, Optimization of a logistics transportation network based on a genetic algorithm. Mob. Inf. Syst. **2022**, 1–8 (2022)
7. X.B. Meng, X.Z. Gao, L. Lu, Y. Liu, H. Zhang, A new bio-inspired optimisation algorithm: bird swarm algorithm. J. Exp. Theor. Artif. Intell. **28**(4), 673–687 (2016)
8. F.A. Hashim, K. Hussain, E.H. Houssein, M.S. Mabrouk, W. Al-Atabany, Archimedes optimization algorithm: a new metaheuristic algorithm for solving optimization problems. Appl. Intell. **51**(3), 1531–1551 (2020)
9. B. Abdollahzadeh, F.S. Gharehchopogh, S. Mirjalili, African vultures optimization algorithm: a new nature-inspired metaheuristic algorithm for global optimization problems. Comput. Ind. Eng. **158**, 107408 (2021)
10. S. Arora, S. Singh, Butterfly optimization algorithm: a novel approach for global optimization. Soft. Comput. **23**(3), 715–734 (2019)
11. F. Martínez-Álvarez et al., Coronavirus optimization algorithm: a bioinspired metaheuristic based on the COVID-19 propagation model. Big Data **8**(4), 308–322 (2020)
12. S. Kaur, L.K. Awasthi, A.L. Sangal, G. Dhiman, Tunicate swarm algorithm: a new bio-inspired based metaheuristic paradigm for global optimization. Eng. Appl. Artif. Intell. **90**, 103541 (2020)
13. L. Abualigah, D. Yousri, M. Abd Elaziz, A.A. Ewees, M.A.A. Al-qaness, A.H. Gandomi, Aquila optimizer: a novel meta-heuristic optimization algorithm. Comput. Ind. Eng. **157**, 107250 (2021)

14. H. Zamani, M.H. Nadimi-Shahraki, A.H. Gandomi, Starling murmuration optimizer: a novel bio-inspired algorithm for global and engineering optimization. Comput. Methods Appl. Mech. Eng. **392**, 114616 (2022)
15. Y. Jiang, Q. Wu, S. Zhu, L. Zhang, Orca predation algorithm: a novel bio-inspired algorithm for global optimization problems. Expert Syst. Appl. **188**, 116026 (2022)
16. M. Dehghani, Z. Montazeri, E. Trojovská, P. Trojovský, Coati optimization algorithm: a new bio-inspired metaheuristic algorithm for solving optimization problems. Knowl.-Based Syst. **259**, 110011 (2023)
17. W. Zhao, L. Wang, S. Mirjalili, Artificial hummingbird algorithm: a new bio-inspired optimizer with its engineering applications. Comput. Methods Appl. Mech. Eng. **388**, 114194 (2022)
18. Y. Yuan et al., Coronavirus mask protection algorithm: a new bio-inspired optimization algorithm and its applications. J. Bionic Eng. **20**(4), 1747–1765 (2023)
19. B. Abdollahzadeh, F. Soleimanian Gharehchopogh, S. Mirjalili, Artificial gorilla troops optimizer: a new nature-inspired metaheuristic algorithm for global optimization problems. Int. J. Intell. Syst. **36**(10), 5887–5958 (2021)
20. A. Ramadan, M. Ebeed, S. Kamel, A.M. Agwa, M. Tostado-véliz, The probabilistic optimal integration of renewable distributed generators considering the time-varying load based on an artificial gorilla troops optimizer. Energies **15**(4), 1302 (2022)
21. M.A. El-Dabah, S. Kamel, M. Khamies, H. Shahinzadeh, G.B. Gharehpetian, Artificial gorilla troops optimizer for optimum tuning of TID based power system stabilizer, in *2022 9th Iranian Joint Congress on Fuzzy and Intelligent Systems (CFIS)* (2022), pp. 1–5
22. M. Abdel-Basset, D. El-Shahat, K.M. Sallam, K. Munasinghe, Parameter extraction of photovoltaic models using a memory-based improved gorilla troops optimizer. Energy Convers. Manag. **252**, 115134 (2022)
23. R.A. Parvin, B. Jana, S. Acharyya, Predicting critical genes from genomic data using artificial gorilla troops optimizer. Easy chair preprints. https://easychair-www.easychair.org/publications/preprint/zd2B. Accessed 23 Apr 2022
24. M.K. Gude, U. Salma, Artificial gorilla troops optimizer for tuning power system stabilizer control parameters, in *2021 IEEE 2nd International Conference on Electrical Power and Energy Systems, ICEPES 2021* (2021)
25. N.A. Baghdadi, A. Malki, H.M. Balaha, M. Badawy, M. Elhosseini, A3 C-TL-GTO: Alzheimer automatic accurate classification using transfer learning and artificial gorilla troops optimizer. Sensors **22**(11), 1–22 (2022)
26. M. Ramesh, A.K. Yadav, P.K. Pathak, Artificial gorilla troops optimizer for frequency regulation of wind contributed microgrid system. J. Comput. Nonlinear Dyn. **18**(1), 1–11 (2022)
27. H. Alsolai et al., Enhanced artificial gorilla troops optimizer based clustering protocol for UAV-assisted intelligent vehicular network. Drones **6**(11), 1–16 (2022)
28. S. Govindaraju, R. Metia, P. Girija, K. Baranitharan, M. Indirani, R. Mahaveerakannan, Detection of DDoS attacks using artificial gorilla troops optimizer based deep learning model, in *Third International Conference on Artificial Intelligence and Smart Energy (ICAIS)* (2023), pp. 385–391
29. M. Fayaz-dastgerdi, H. Shahinzadeh, J. Moradi, H. Nafisi, A. Karimi, G. B. Gharehpetian, Optimal power flow in an islanded renewable microgrid using artificial gorilla troops optimizer, in *5th International Conference on Optimizing Electrical Energy Consumption (OEEC)* (2023), pp. 20–26
30. M.F. Isham et al., Bearing fault diagnosis using extreme learning machine based on artificial gorilla troops optimizer, in *Advances in Intelligent Manufacturing and Mechatronics* (Springer Nature Singapore, Singapore, 2023), pp. 87–103
31. V. Plevris, G. Solorzano, A collection of 30 multidimensional functions for global optimization benchmarking. Data (Basel) **7**(4), 1–51 (2022)
32. K. Tai, A.-R. El-Sayed, M. Biglarbegian, C.I. Gonzalez, O. Castillo, S. Mahmud, Review of recent type-2 fuzzy controller applications. Algorithms **9**(2), 39 (2016)
33. P. Melin, O. Castillo, A new method for adaptive control of non-linear plants using type-2 fuzzy logic and neural networks. Int. J. Gen. Syst. **33**(2–3), 289–304 (2004)

34. P. Melin, O. Castillo, Adaptive intelligent control of aircraft systems with a hybrid approach combining neural networks, fuzzy logic and fractal theory. Appl. Soft Comput. **3**(4), 353–362 (2003)
35. E. Ontiveros, P. Melin, O. Castillo, Comparative study of interval type-2 and general type-2 fuzzy systems in medical diagnosis. Inf. Sci. **525**, 37–53 (2020)
36. J.E. Moreno, M.A. Sanchez, O. Mendoza, A. Rodriguez-Diaz, O. Castillo, P. Melin, J.R. Castro, Design of an interval type-2 fuzzy model with justifiable uncertainty. Inf. Sci. **513**, 206–221 (2020)
37. F. Valdez, J.C. Vazquez, P. Melin, O. Castillo, Comparative study of the use of fuzzy logic in improving particle swarm optimization variants for mathematical functions using co-evolution. Appl. Soft Comput. **52**, 1070–1083 (2017)
38. D. Sanchez, P.M elin, O. Castillo, A grey wolf optimizer for modular granular neural networks for human recognition. Comput. Intell. Neurosci. **2017** (2017). https://doi.org/10.1155/2017/4180510
39. O. Castillo, E. Lizzarraga, J. Soria, P. Melin, F. Valdez, New approach using ant colony optimization with ant set partition for fuzzy control design applied to the ball and beam system. Inf. Sci. **294**, 203–215 (2015)
40. L. Amador-Angulo, O. Mendoza, J.R. Castro, A. Rodriguez-Diaz, P. Melin, O. Castillo, Fuzzy sets in dynamic adaptation of parameters of a bee colony optimization for controlling the trajectory of an autonomous mobile robot. Sensors **16**(9), 1458 (2016)

A Hybrid Fuzzy Fireworks and Grey Wolf Metaheuristic Algorithm

Juan Barraza, Luis Rodriguez, Fevrier Valdez, Patricia Melin, and Oscar Castillo

1 Introduction

Computational Intelligence focuses on the study the adaptive mechanism to allow the performance of the complex and changing systems, which imitate the nature and human linguistic reasoning. We can classify in three different pillars: artificial neural networks [1], fuzzy logic [2] and evolutionary computation [3, 4].

For this investigation, we focus on the evolutionary computing and fuzzy logic; as we know into the evolutionary computing there are branches, one of some branches is the swarm intelligence [5, 6], which is conformed for a wide variety of the optimization algorithms [7, 8], such as Fireworks Algorithm [9–11] and Grey Wolf Optimizer [12].

Continuing on the field of intelligent computing [13, 14], fuzzy logic has been implemented by scientific researchers in different applications [15–21]. Therefore, we take the examples of previous works [22–24] to dynamically adjust the parameter in the hybrid algorithm [25].

J. Barraza · L. Rodriguez · F. Valdez · P. Melin · O. Castillo (✉)
Tijuana Institute of Technology, TecNM, Tijuana, Mexico
e-mail: ocastillo@tectijuana.mx

F. Valdez
e-mail: fevrier@tectijuana.mx

P. Melin
e-mail: pmelin@tectijuana.mx

P. Melin and O. Castillo (eds.), *New Directions on Hybrid Intelligent Systems Based on Neural Networks, Fuzzy Logic, and Optimization Algorithms*, Studies in Computational Intelligence 1146, https://doi.org/10.1007/978-3-031-53713-4_16

Hybridization is a word that has gained strength in the field of computer sciences. Which refer to combined two or more similar or different computing systems [26, 27]. For example, we could combine two kinds of artificial neural networks, combine artificial neural network with fuzzy logic [20, 28, 29], or, as this case, we combined two optimization algorithms [25], FWA and GWO.

The main contribution of this method is to show the importance of using Fuzzy Logic into the optimization algorithms. For this work, we are improving the performance of FWA-GWA, introducing a Fuzzy Inference System with the objective to dynamically adjustment the amplitude coefficient [30], and then, we achieved an equilibrium between the exploration and exploitation [31–33] of the Hybrid Algorithm.

This paper is organized as follows: Sect. 2 reviews the Conventional FWA-GWO Hybrid Algorithm [25]. In Sect. 3, the proposed method (F-FWA-GWO) is presented. Section 4 shows the experiments and results; Sect. 5 outlines the conclusions.

2 Conventional Hybrid Algorithm (FWA-GWO)

The hybridization between these two algorithms, called FWA-GWO is presented in the article "A new hybridization Approach between the Fireworks Algorithm and Grey Wolf Algorithm" in 2018 [25] where the authors take advantages of the originals optimization algorithms Fireworks Algorithm [34, 35] by Tan and its features as (initialization, explosion amplitude and locations) and the Grey Wolf Algorithm [12] by Mirjalili and the features like hierarchical pyramid, interaction among the population and the next location of population.

2.1 Advantages of FWA-GWO Algorithm

In Fig. 1 we can find the general flow chart of the FWA-GWO optimization algorithm.

Where it is important mention that the red color in the flow chart represents behavior of the Fireworks algorithm and the blue steps are features of the Grey Wolf Optimizer.

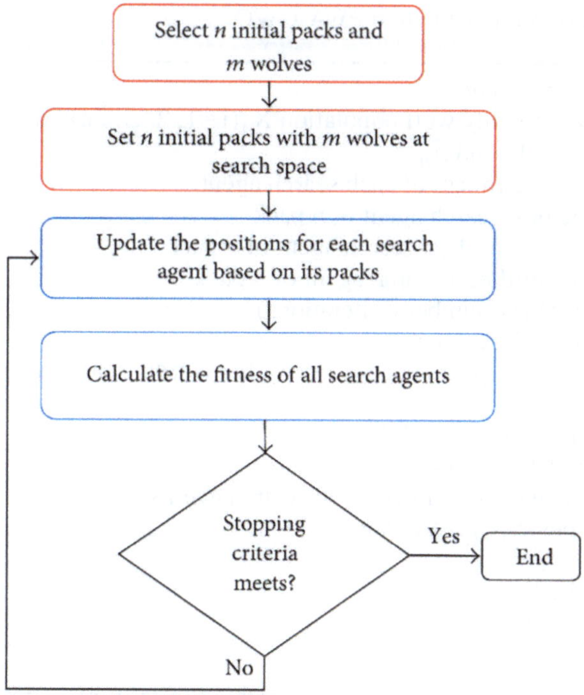

Fig. 1 Flow chart of the FWA-GWO [25]

In addition, we can find the Pseudocode for the FWA-GWO algorithm in the following description of Algorithm 1.

Algorithm 1 Conventional hybrid FWA-GWO

```
Initialize n and m
Initialize the grey wolf population X_ni (i= 1, 2, ... , n)
Initialize a, E_n and Â_n
Calculate the fitness of each search agent
X_nα = the best search agent of n pack
X_nβ = the second-best search agent of n pack
X_nδ = the third best search agent of n pack
while (t <Max number of iterations)
      for each search agent
              Update the position of the current search agent
      end for
      Update a, Â_n and E_n
      Update n and m
      Calculate the fitness of all search agents
      Update X_nα, X_nβ and X_nδ
      t = t + 1
end while
return best (X_1α, X_2α, ... , X_nα)
```

It is important to mention that in the proposed method, we can work with the updates, specifically in the A_n value, where main equation can be described as a

$$A_i = \widehat{A}.\frac{f(x_i) - y_{min} + \epsilon}{\sum_{i=1}^{n}(f(x_i) - y_{min}) + \epsilon} \tag{1}$$

In order to apply a technique of computational intelligence, we are proposing a method that is described in the next section of this paper.

3 Proposed Method (F1FWA-GWO)

In this Section, we explain in detail the modification of the conventional Hybrid Algorithm (FWA-GWO).

We implemented the Hybrid Algorithm (FWA-GWO) with an adjustment of parameter (Amplitude coefficient) using a Fuzzy Logic Inference System to automatically find the decrement equilibrium between exploration and exploitation into this Hybrid Algorithm. We decided to call this as Hybrid Fuzzy Fireworks and Grey Wolf Algorithm and we denoted as F-FWAGWO as mentioned in previous sections.

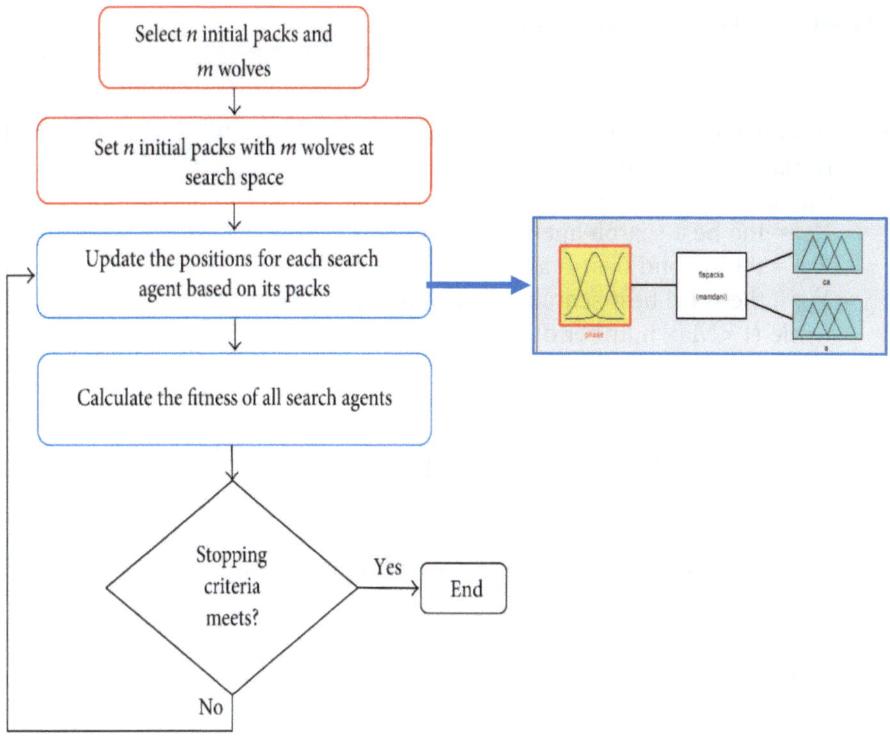

Fig. 2 Flow chart of the F1FWA-GWO

The proposed F-FWA-GWO method has four steps:

1. Select n initial locations and m wolves.
2. Set n initial packs with m wolves at search space.
3. Update the positions for each search agents based on its packs.
    - Dynamically adjustment parameter with Fuzzy Inference System.
4. Calculate the fitness of all search agents.

The steps are described in more detail below in Algorithm 2, but before we are presenting these steps in a graphic way, in the following Fig. 2.

Algorithm 2 Fuzzy hybrid FWA-GWO

```
Initialize n and m
Initialize the grey wolf population X_ni (i= 1, 2, … , n)
Initialize a, E_n and Â_n
Calculate the fitness of each search agent
X_nα = the best search agent of n pack
X_nβ = the second best search agent of n pack
X_nδ = the third best search agent of n pack
while (t <Max number of iterations)
        for each search agent
                Update the position of the current search agent
        end for
        Update HFA^ (using fuzzy logic)
        Update a and E_n
        Update n and m
        Calculate the fitness of all search agents
        Update X_nα, X_nβ and X_nδ
        t = t + 1
end while
return best (X_1α, X_2α, … , X_nα)
```

As we mentioned before, this work is a variant of a previous work named as A New Hybridization Approach between the Fireworks Algorithm and Grey Wolf Optimizer Algorithm, and we called as Conventional Hybrid Algorithm, therefore, the steps 1, 2 and 4 are explained a detail in [25].

In this section, we focus on the Step 3 to explaining the dynamically adjustment of parameter with Fuzzy Inference System. The parameter that modify into the Conventional Hybrid Algorithm is Amplitude Coefficient, which is the main parameter to controlling the explosion amplitude of the agents based on its packs. We are using the fuzzy adjustment to have a balance between exploration and exploitation, respectively.

The Equation modified in Eq. 2 is shown:

$$HFA_i = \widehat{HFA}.\frac{f(x_i) - y_{min} + \epsilon}{\sum_{i=1}^{n}(f(x_i) - y_{min}) + \epsilon} \tag{2}$$

where $\widehat{HFA}$ will be in a range between 0 and 40.

Figure 3 shows the Fuzzy Inference System that was used.

As we can see, the Fuzzy Inference System is formed by one input variable and two output variables and, for this variation of the FFWA-GWO, we just to use Triangular membership functions.

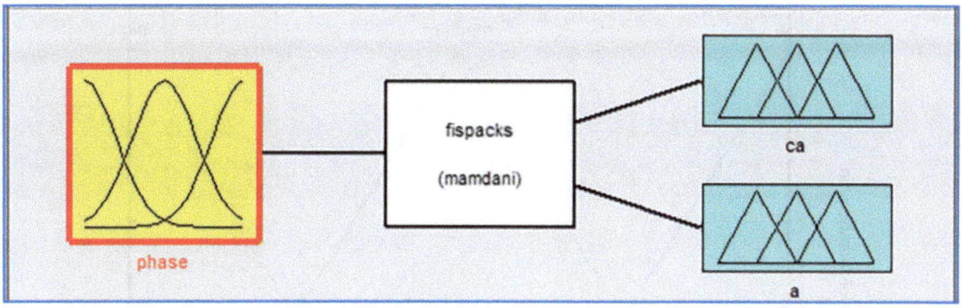

Fig. 3 Fuzzy inference system for F1FWA-GWO

Figure 4 shows the input variable and output variables are shown in Figs. 5 and 6, respectively.

Phase is a variable with a range between 0 and 1. Input variable is denoted as phase and it has three partitions: low [−0.6, 0, 0.4] medium [0.1, 0.5, 0.9] and high [0.6, 1, 1.4].

The Fuzzy rules are shown in Fig. 7.

To control dynamically *ca* and *a* as amplitude parameter, we used the fuzzy rules, where ca decrease 40 to 0 and a decrease 2 to 0.

To verify that the dynamically adjustment exists, we added the performance surface of ca parameter, which is shown in Fig. 8.

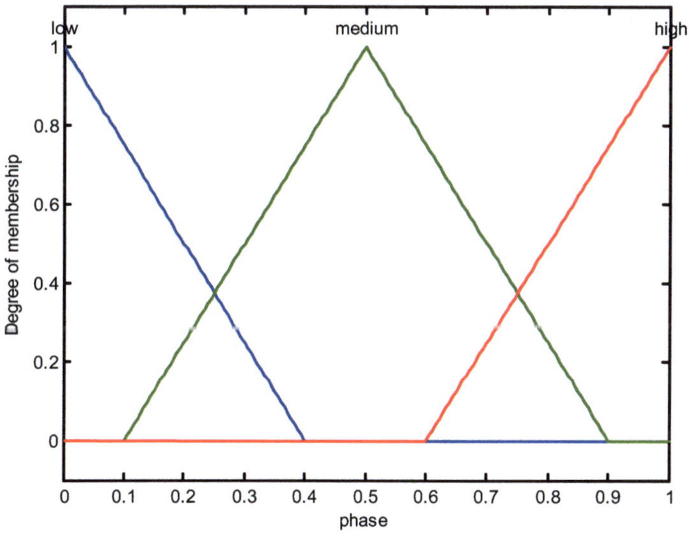

Fig. 4 Input variable for fuzzy inference system

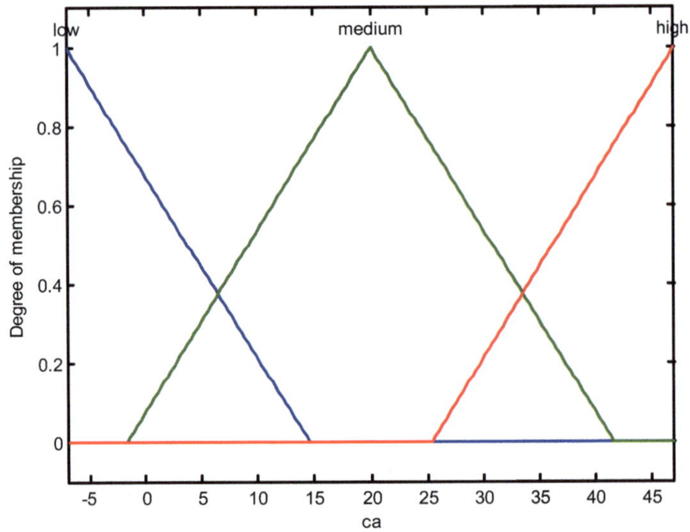

Fig. 5 Output variable (ca) for fuzzy inference system

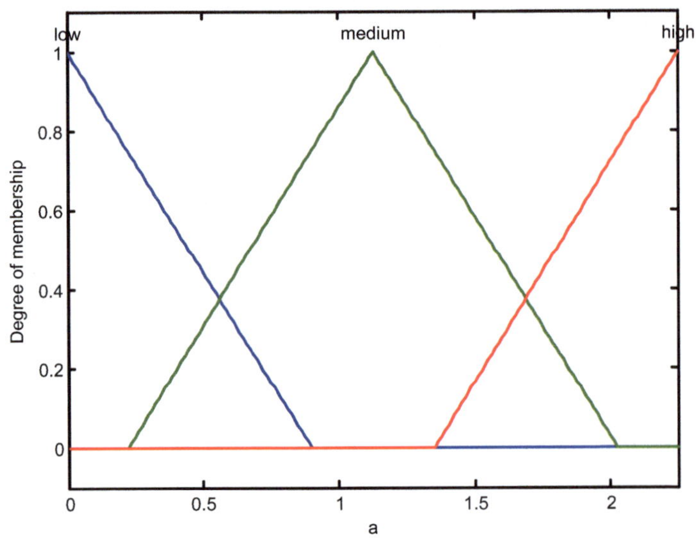

Fig. 6 Output variable (a) for fuzzy inference system

1. If (phase is low) then (ca is high)(a is high) (1)
2. If (phase is medium) then (ca is medium)(a is medium) (1)
3. If (phase is high) then (ca is low)(a is low) (1)

Fig. 7 Fuzzy rules for fuzzy inference system

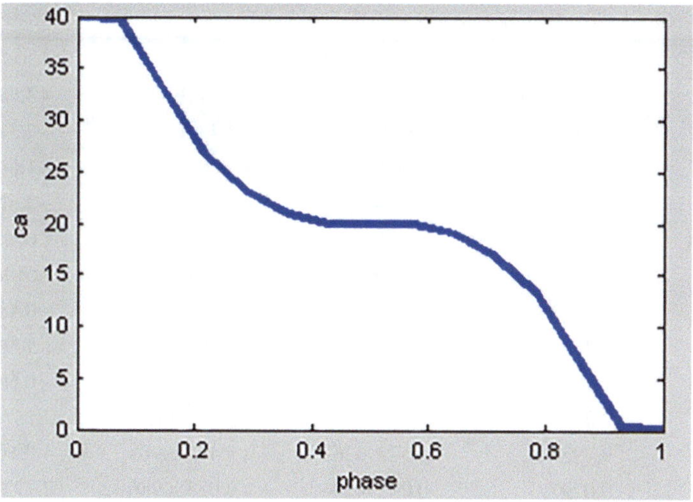

Fig. 8 Convergence of the ca parameter

4 Experiments and Results

In this section we present the results obtained by the experiments realized with the fuzzy logic in the hybrid method.

It is important to mention that the results are compared with statistical test (z test) where we obtained mean and standard deviation of 31 independent experiments for each benchmark functions.

In Tables 1, 2, and 3 we can see the z-test results where the functions have 30 dimensions respectively and the versions that the authors mentioned in the original paper [25].

In these Tables 1, 2, and 3 we can see that the fuzzy logic has better performance in the version 3, also it is important mention that according of the experiments, the fuzzy logic helps to improve the results where the problem has more dimensions.

In Tables 4, 5, and 6 we can note that the proposed method improved the performance of the hybrid FWA-GWO, as Table 6 shows, because the proposed method is better in 8 of 13 benchmark function that was analyzed.

Finally, in Tables 7, 8, and 9 we can note the proposed method where the problem has 90 dimensions and we can see that the fuzzy logic improves the results where the problem is more complex.

Table 1 Version 1 with 30 dimensions

FWA-GWO		F-FWA-GWO		Z-test
6.43E−28	1.17E−27	3.7917E−31	6.444E−31	**−3.90813**
1.70E−16	1.63E−16	2.0619E−18	2.482E−18	**−7.37330**
8.47E−04	0.0025	0.000255	0.0004721	**−1.64506**
0.0383	0.0584	0.014492	0.0274842	**−2.62962**
27.6283	0.5878	27.83115	0.6177712	1.69880
0.8856	0.4284	1.030633	0.4749305	1.61986
775.8908	98.3576	766.2052	102.77187	−0.48623
1.7697	6.7631	33.7826588	174.493368	1.30920
9641.7985	11406.7060	8512.894	8726.36565	−0.56134
1.7310	2.2942	3.16412801	4.07656679	2.18786
5.8606	9.4934	5.57996263	9.41156908	−0.14993
0.0693	0.0297	0.09304294	0.05350986	2.77379
0.00	0.00	0	0	N/A

Table 2 Version 2 with 30 dimensions

FWA-GWO		F-FWA-GWO		Z-test
1.8439E−22	7.4937E−22	1.2294E−25	2.7776E−25	**−1.75599**
1.1714E−13	1.2045E−13	2.2062E−15	3.1455E−15	**−6.81195**
0.00436445	0.01660925	0.00105821	0.00169155	−1.41425
0.04087664	0.0440062	0.01227169	0.01133333	**−4.4953**
27.5221951	0.52759271	27.6249485	0.57155474	0.94339
0.57463951	0.3179441	0.77352824	0.42940365	2.65833
811.883332	98.7647041	865.248978	120.539596	2.44559
7.5452E−05	0.00019101	8.7058E−05	0.0001342	0.35505
15241.4387	16021.8731	15241.7609	16459.6479	0.00010
1.3627	2.04846158	3.42091013	5.0312965	2.70573
1.7074E−11	4.2341E−11	2.9434E−13	2.386E−13	**−2.8300**
0.07051915	0.0316306	0.08847912	0.04311549	2.39855
0	0	0	0	N/A

5 Conclusions

The conclusions of this work could be described as follows: we have included fuzzy logic in the hybrid optimization algorithm as a test of optimization proposal.

The adjustment of amplitude parameter is based on previous works [22, 23, 25]; therefore, we can decided to start adjustment the parameter mentioned above to main

Table 3 Version 3 with 30 dimensions

FWA-GWO		F-FWA-GWO		Z-test
1.7805E−18	2.697E−18	7.4421E−21	2.6012E−20	**−4.69452**
4.1937E−11	3.5187E−11	7.1769E−13	9.6217E−13	**−8.36261**
0.00948489	0.02410614	0.00272137	0.00585984	**−1.94699**
0.02480709	0.02068105	0.00611933	0.00726259	**−6.08860**
27.7053665	0.58471148	27.6817202	0.63107266	−0.19628
0.59296838	0.35280547	0.58170233	0.33792797	−0.16468
−753.73446	291.269306	−1019.0747	100.421428	**−6.15041**
0.15439472	0.69042326	0.00299533	0.00627116	−1.56594
6439.17742	23019.3982	4023.09589	7454.00764	−0.71309
3.10659088	3.92435119	7.81481201	16.5208143	1.98011
1.3312E−09	1.3413E−09	5.4704E−11	7.488E−11	**−6.78555**
0.04158109	0.02490691	0.03290371	0.03142038	−1.54555
9.3259E−17	2.2132E−16	0	0	**−3.00924**

Table 4 Version 1 with 60 dimensions

FWA-GWO		F-FWA-GWO		Z-test
6.78E−18	1.14E−17	5.646E−20	9.7462E−20	**−4.19362019**
1.55E−11	1.28E−11	7.8113E−13	6.0518E−13	**−8.23001623**
2.3311	2.0824	3.69503992	6.29842082	1.46829282
3.2579	1.1983	1.94567547	0.93937705	**−6.15455308**
57.9631	0.7311	58.4113506	0.42496459	3.78502031
4.3132	0.7794	4.79576959	1.11194124	2.5382043
1273.7126	157.2159	1439.07009	276.638796	3.71125149
975.1898	2020.46	1550.89379	3471.33845	1.02360868
123474.2900	105804.6016	103240.647	93645.9949	−1.02266492
5.31247553	6.25638053	2.2737E−13	6.7855367	**−4.11053559**
13.3142	10.0509	12.0732227	10.528144	−0.60885888
0.1571	0.1423	0.13765517	0.11911	−0.74822201
1.04E−16	3.11E−16	0	0	**−2.39542637**

goal of controlling the balance between exploration and explotation into the Hybrid Algorithm.

The conclusions are divided according to the three versions of the hybrid algorithm:

Table 5 Version 2 with 60 dimensions

FWA-GWO		F-FWA-GWO		Z-test
4.6071E−14	1.3052E−13	4.2928E−16	5.0004E−16	**−2.49738444**
2.3724E−09	1.3658E−09	1.8674E−10	1.0544E−10	**−11.3942863**
13.7898346	15.3383444	11.2738178	20.4171743	−0.7036115
1.12511368	0.34685491	0.66698753	0.24647828	**−7.688809**
57.9556247	0.64565014	58.1378675	0.65421986	1.41592758
3.8645452	0.79177514	4.54580227	0.98721447	3.84443571
1572.60172	164.594448	1501.36629	218.129785	**−1.86166891**
1.93907187	13.6862139	0.00048153	0.00045033	−1.01155106
97623.3189	73980.7882	191977.714	248788.34	2.59607856
6.84271011	5.16272345	17.7804303	11.4381487	6.2243264
3.6804E−07	1.2271E−06	4.2744E−08	8.2434E−08	**−1.88879847**
0.56605589	0.41456579	0.41795133	0.28060262	**−2.11281131**
1.0671E−14	2.4179E−14	1.6283E−16	3.5197E−16	**−3.10344338**

Table 6 Version 3 with 60 dimensions

FWA-GWO		F-FWA-GWO		Z-test
8.1705E−11	1.9133E−10	7.2833E−13	9.737E−13	**−3.02248409**
1.2425E−07	1.0068E−07	8.4237E−09	4.4214E−09	**−8.20745086**
3.96142545	5.02258205	3.33377199	7.18451317	−0.51133
1.23165872	0.64112394	0.58490902	0.3583958	**−6.28826234**
58.5642592	0.5193704	58.0825728	0.70066279	**−3.94412131**
4.61163933	1.05648508	3.81923484	1.36265424	**−3.28197664**
2020.51853	251.665096	2051.93557	140.466444	0.77846352
81.8637242	197.113652	13.0644383	27.3261457	**−2.46898601**
231548.526	432289.986	245848.064	469573.779	0.15999644
21.6702093	30.8963329	26.2850703	39.2446615	0.65983024
5.1019E−06	9.2377E−06	3.4722E−07	3.3752E−07	**−3.67325678**
0.07230841	0.05656685	0.06747057	0.04515742	−0.47732311
3.145E−11	1.3113E−10	4.4305E−14	7.8978E−14	**−1.71042654**

1. For Version 1 of the Fuzzy Hybrid Algorithm (F-FWA-GWO), this improves the results in 4 of 12 (33%) benchmark funtions with 30 dimensions; For 60 dimension the Fuzzy Hybrid Algorithm won in 5 of 12 (42%) benchmarck functions and, for 90 dimensions 7 of 12 (67%).

Table 7 Version 1 with 90 dimensions

FWA-GWO		F-FWA-GWO		Z-test
3.59E−13	1.05E−12	3.2516E−15	4.7224E−15	**−2.42453979**
1.71E−09	9.38E−10	1.8024E−10	1.4836E−10	**−11.4811895**
170.1456	208.2290	193.618981	299.3565	0.45970369
9.7444	3.1814	7.95980326	2.75127298	**−3.03004695**
88.8722	0.4571	88.5873508	0.51630934	**−2.95013805**
8.8690	1.4223	9.29971666	1.6782257	1.39828772
2958.5273	123.5623	2703.64595	221.202277	**−7.18392957**
32391.5762	49298.17	36416.297	49526.723	0.41130912
427794.0870	576474.7710	232604.291	207210.617	**−2.27549774**
19.1615	20.6260	19.7532278	12.859161	0.1738474
14.8326	9.8082	15.5387448	9.53044015	0.36873042
0.4483	0.5499	0.08318967	0.19102033	**−4.4787198**
6.11E−12	3.76E−11	2.4292E−13	9.9224E−13	−1.11282233

Table 8 Version 2 with 90 dimensions

FWA-GWO		F-FWA-GWO		Z-test
3.178E−10	7.6849E−10	9.7734E−12	1.0828E−11	**−2.86212252**
1.4535E−07	1.0176E−07	2.0059E−08	1.2674E−08	**−8.72510186**
218.370239	141.990755	286.673605	537.780698	0.87697756
3.20324609	0.95509621	2.10799584	0.63951277	**−6.80481893**
89.5622236	5.10369477	88.4208582	0.367222	−1.59295616
9.03801979	1.23076665	9.95651728	1.37970424	3.54775404
2382.76546	133.511104	2347.8836	189.400671	−1.07499481
1.33402153	5.11179162	0.01684986	0.05642576	**−1.84004257**
353505.705	442838.277	244463.048	227944.095	−1.56350554
28.2039797	20.3174953	39.7093502	16.2966578	3.15463157
0.00020097	0.00045043	1.3775E−05	2.4062E−05	**−2.96368319**
1.36414993	0.57788807	0.81418564	0.44582029	**−5.38113656**
1.2064E−09	4.3776E−09	0.00181529	0.00994275	1.30383962

2. With Version 2 of the Fuzzy Hybrid Algorithm and using 30 dimension in the benchmarck functions, the results show a 33% improvement, i.e., 4 of 12 functions, for 60 dimension the improvement is 58% (7 of 12 functions) and for 90 dimensions is 50% (6 of 12 funtions).
3. Version 3 of the Fuzzy Hybrid Algorithm shows better results than Version 1 and 2, i.e., for 30 dimensions F1FWA-GWO was obtained best results in 7 of 12 functions (58%), in 60 dimensions F1FWA-GWO won in 8 of 12 (67%)

Table 9 Version 3 with 90 dimensions

FWA-GWO		F-FWA-GWO		Z-test
6.0184E−07	9.5828E−07	1.5279E−08	3.6002E−08	**−4.36819439**
4.3474E−06	2.2746E−06	4.0567E−07	2.6027E−07	**−12.2952117**
68.9997491	71.6441609	60.5302292	76.4441731	−0.57731124
3.68783247	1.66068285	3.0719967	1.22398164	**−2.13181379**
90.1465212	1.46842867	88.648169	0.66426141	**−6.63925012**
7.54375831	2.50300462	5.53516021	1.88276363	**−4.57980905**
−2485.26695	427.571804	−2322.44754	184.793618	2.49629028
397.220794	835.350928	302.334344	342.989232	−0.7503949
541955.021	846552.816	978250.881	1535745.54	1.77677279
64.8371216	63.4858787	73.7948615	86.831887	0.59471991
0.00031502	0.00042725	2.9247E−05	2.2495E−05	**−4.77002315**
0.24547352	0.7293556	0.30429546	0.42637097	0.4972226
8.5957E−07	4.9515E−06	2.5403E−09	7.8419E−09	−1.2360742

benchmarck functions and, finally, for 90 dimensions F-FWA-GWO won in 6 of 12 benchmark funtcions (50%).

A general way to conclude this study, it is to saying that the better perormance of the Fuzzy Hybrid Algorithm is with a high number of dimensions, referensing to a complex problems. The results show a percent average improvement of 56% for 60 and 90 dimensions, on the contrary, for 30 dimension only the 42% of improvement.

For future works, firstly, we could test the adjustment of another parameter in the Fuzzy Hybrid Algorithm (F-FWA-GWO), for example, the parameter could be the spark coefficient [36], C parameter, among others. We also can test with variations of the membership functions of the Fuzzy Inference System; the variations could be using to Gaussian and Trapezoidal membership functions. A second option could be to apply Interval Type-2 Fuzzy Logic [28] in dynamical adjustment of parameters, like in [37–39]. Also, we could apply the optimization method in other kinds of problems, such as in [40–44]. Finally, a more systematic comparison of results could be done with other algorithms, like [45–47].

References

1. C. Chen, L. Mi, Z. Liu, B. Qiu, H. Zhao, L. Xu, Predefined-time synchronization of competitive neural networks. Neural Netw. **142**, 492–499 (2021)
2. L.A. Zadeh, Knowledge representation in fuzzy logic. IEEE Trans. Knowl. Data Eng. **I**(I), 89-0084 (1989)
3. F. Aladwan, M. Alshraideh, M. Rasol, A genetic algorithm approach for breaking of simplified data encryption standard. Int. J. Secur. Appl. **9**(9), 295–304 (2015)

4. P. Nerurkar, A. Shirkeb, M. Chandanec, S. Bhirudd, A novel heuristic for evolutionary clustering. Procedia Comput. Sci. **125**, 780–789 (2018)
5. E. Bonabeau, M. Dorigo, G. Theraulaz, *Swarm Intelligence: From Natural to Artificial Systems* (OUP USA, 1999)
6. B. Melián, J. Moreno, Metaheuristics: a global vision. Ibero-Am. J. Artif. Intell. **19**, 7–28 (2003)
7. U. Can, B. Alatas, Physics based metaheuristic algorithms for global optimization. Am. J. Inf. Sci. Comput. Eng. **1**, 94–106 (2015)
8. D.H. Wolpert, W.G. Macready, No free lunch theorems for optimization, in *Evolut. Comput., IEEE Trans* (1997), pp. 67–82
9. J. Li, S. Zheng, Adaptive fireworks algorithm, in *IEEE Congress on Evolutionary Computation (CEC)* (2014), pp. 3214–3221
10. Y. Tan, *Fireworks Algorithm* (Springer-Verlag, Berlin Heidelberg, 2015), pp.355–364
11. Y. Tan, Y. Zhu, *Fireworks Algorithm for Optimization* (Springer-Verlag, Berlin Heidelberg, 2010), pp.355–364
12. L. Rodriguez, O. Castillo, J. Soria, Grey wolf optimizer (GWO) with dynamic adaptation of parameters, in *IEEE CEC 2016* (2016), pp. 3116–3123
13. L. Telescaa, M. Bernardib, C. Rovellib, Intra-cluster and inter-cluster time correlations in lightning sequences. Physica A **356**, 655–661 (2005)
14. B.Y. Wu, On the intercluster distance of a tree metric. Theoret. Comput. Sci. **369**, 136–141 (2006)
15. N.H. Abdulmajeed, M. Ayob, A firework algorithm for solving capacitated vehicle routing problem. Int. J. Adv. Comput. Technol. (IJACT) **6**(1), 79–86 (2014)
16. X. Chena, S. Liua, T. Chena, Z. Zhangb, H. Zhangb, An improved semi-supervised clustering algorithm for multi-density datasets with fewer constraints. Procedia Eng. **29**, 4325–4329 (2012)
17. M. Simoes, K. Bose, J. Spiegel, Fuzzy logic based intelligent control of a variable speed cage machine wind generation system. IEEE Trans. Power Electron. **12**(1), 87–95 (1997)
18. M.A. Sanchez, O. Castillo, J.R. Castro, P. Melin, Fuzzy granular gravitational clustering algorithm for multivariate data. Inf. Sci. **279**, 498–511 (2014)
19. Y. Zheng, Q. Song, S.-Y Chen, Multiobjective fireworks optimization for variable-rate fertilization in oil crop production. Appl. Soft Comput. **13**, 4253–4263 (2013)
20. D. Sanchez, P. Melin, O. Castillo, A grey wolf optimizer for modular granular neural networks for human recognition. Comput. Intell. Neurosci. **2017**, 4180510:1–4180510:26 (2017)
21. X. Liu, X. Qin, A neighborhood information utilization fireworks algorithm and its application to traffic flow prediction. Expert Syst. Appl. **183**, 115189 (2021)
22. J. Barraza, P. Melin, F.Valdez, Fuzzy FWA with dynamic adaptation of parameters, in *IEEE CEC 2016*, pp. 4053–4060
23. J. Barraza, F. Valdez, P. Melin, C. Gonzalez, Fireworks algorithm (FWA) with adaptation of parameters using fuzzy logic, in *Nature-Inspired Design of Hybrid Intelligent Systems* (2017), pp. 313–327
24. L. Rodríguez, O. Castillo, J. Soria, A study of parameters of the grey wolf optimizer algorithm for dynamic adaptation with fuzzy logic, in *Nature-Inspired Design of Hybrid Intelligent Systems* (2017), pp. 371–390
25. J. Barraza, L. Rodríguez, O. Castillo, P. Melin, F. Valdez, A new hybridization approach between the fireworks algorithm and grey wolf optimizer algorithm. J. Optim. Res. Artic. **2018**, 6495362 (2018)
26. J. Soler, F. Tencé, L. Gaubert, C. Buche, Data clustering and similarity, in *Proceedings of the Twenty-Sixth International Florida Artificial Intelligence Research Society Conference* (2013), pp. 492–495
27. H.A. Sturges, The choice of a class interval. J. Am. Stat. Assoc. **21**(153), 65–66 (1926)
28. E. Rubio, O. Castillo, Interval type-2 fuzzy possibilistic C-means optimization using particle swarm optimization, in *Nature-Inspired Design of Hybrid Intelligent Systems* (2017), pp. 63–78

29. J. Soto, P. Melin, Optimization of the fuzzy integrators in ensembles of ANFIS model for time series prediction: the case of Mackey-Glass, in *IFSA-EUSFLAT 2015*
30. J. Barraza, P. Melin, F. Valdez, C.I. González, Fuzzy fireworks algorithm based on a sparks dispersion measure. Algorithms **10**(3), 83 (2017)
31. R. Larson, B. Farber, *Elementary Statistics Picturing the World* (Pearson Education Inc., 2003), pp. 428–433
32. M. Črepinšek, S.H. Liu, M. Mernik, Exploration and exploitation in evolutionary algorithms: a survey. ACM Comput. Surv. **45**(3), 35:32 (2013)
33. J. Liu, S. Zheng, Y. Tan, The improvement on controlling exploration and exploitation of firework algorithm, in *Advances in Swarm Intelligence* (Springer, 2013), pp. 11–23
34. Y. Tan, S. Zheng, Dynamic search in fireworks algorithm, in *Evolutionary Computation (CEC 2014)*
35. J. Li, Y. Tan, A comprehensive review of the fireworks algorithm. ACM Comput. Surv. (CSUR) **52**(6), 1–28 (2019)
36. J. Barraza, P. Melin, F. Valdez, C.I. González, O. Castillo, Iterative fireworks algorithm with fuzzy coefficients, in *FUZZ-IEEE 2017*, Naples, Italy, 9–12 July 2017, pp. 1–6
37. F. Olivas, F. Valdez, P. Melin, A. Sombra, O. Castillo, Interval type-2 fuzzy logic for dynamic parameter adaptation in a modified gravitational search algorithm. Inf. Sci. **476**, 159–175 (2019)
38. F. Olivas, F. Valdez, O. Castillo, P. Melin, Dynamic parameter adaptation in particle swarm optimization using interval type-2 fuzzy logic. Soft. Comput. **20**(3), 1057–1070 (2016)
39. F. Olivas, F. Valdez, O. Castillo, C.I. Gonzalez, G. Martinez, P. Melin, Ant colony optimization with dynamic parameter adaptation based on interval type-2 fuzzy logic systems. Appl. Soft Comput. **53**, 74–87 (2017)
40. E. Ontiveros, P. Melin, O. Castillo, Comparative study of interval type-2 and general type-2 fuzzy systems in medical diagnosis. Inf. Sci. **525**, 37–53 (2020)
41. J.E. Moreno, M.A. Sanchez, O. Mendoza, A. Rodriguez-Diaz, O. Castillo, P. Melin, J.R. Castro, Design of an interval type-2 fuzzy model with justifiable uncertainty. Inf. Sci. **513**, 206–221 (2020)
42. K. Tai, A.-R. El-Sayed, M. Biglarbegian, C.I. Gonzalez, O. Castillo, S. Mahmud, Review of recent type-2 fuzzy controller applications. Algorithms 9(2), 39 (2016)
43. O. Castillo, E. Lizzarraga, J. Soria, P. Melin, F. Valdez, New approach using ant colony optimization with ant set partition for fuzzy control design applied to the ball and beam system. Inf. Sci. **294**, 203–215 (2015)
44. L. Amador-Angulo, O. Mendoza, J.R. Castro, A. Rodriguez-Diaz, P. Melin, O. Castillo, Fuzzy sets in dynamic adaptation of parameters of a bee colony optimization for controlling the trajectory of an autonomous mobile robot. Sensors **16**(9), 1458 (2016)
45. F. Valdez, J.C. Vazquez, P. Melin, O. Castillo, Comparative study of the use of fuzzy logic in improving particle swarm optimization variants for mathematical functions using co-evolution. Appl. Soft Comput. **52**, 1070–1083 (2017)
46. M. Guerrero, F. Valdez, O. Castillo, Comparative study between type-1 and interval type-2 fuzzy systems in parameter adaptation for the cuckoo search algorithm. Symmetry **14**, 2289 (2022). https://doi.org/10.3390/SYM14112289
47. F. Cuevas, O. Castillo, P. Cortés-Antonio, Generalized type-2 fuzzy parameter adaptation in the marine predator algorithm for fuzzy controller parameterization in mobile robots. Symmetry **14**, 859 (2022). https://doi.org/10.3390/SYM14050859

FSC
www.fsc.org
MIX
Papier aus verantwortungsvollen Quellen
Paper from responsible sources
FSC® C105338

If you have any concerns about our products,
you can contact us on
ProductSafety@springernature.com

In case Publisher is established outside the EU,
the EU authorized representative is:
Springer Nature Customer Service Center GmbH
Europaplatz 3, 69115 Heidelberg, Germany

Printed by Libri Plureos GmbH
in Hamburg, Germany